Optoelectronics

By Vaughn D. Martin

Optoelectronics

By Vaughn D. Martin

VOLUME 1
The Introduction
A Self-Teaching Text, Including:
- **Basic Concepts**
- **Photometrics**
- **Optics**

A Division of Howard W. Sams & Company
A Bell Atlantic Company
Indianapolis, IN

International Standard Book Number: 0-7906-1091-4

LIBRARY OF CONGRESS CATALOG CARD NUMBER: 96-72184

Acquisitions Editor: Candace M. Hall
Editor: Natalie F. Harris
Assistant Editors: Pat Brady, Loretta Leisure, Karen Mittelstadt
Typesetting: Natalie Harris
Indexing: Loretta Leisure
Pasteup: Debra Delk, Suzanne Lincoln, Christy Pierce, Scott Stadler, Kelli Ternet, Phil Velikan
Cover Design: Kelli Ternet
Graphics Conversion: Debra Delk, Scott Stadler, Terry Varvel
Illustrations and Other Materials: Courtesy of 3M Corporation, Centronic, Data Display Products, Dialight, EG&G Optoelectronics, ETI Inc., Gilway Technical Lamp, Hamamatsu, Hewlett-Packard, International Light, Irvine Sensors, Keithley Instruments, Peter A. Keller, Miltec Corporation, Motorola, Raytek, SUNX, Tektronix, Texas Instruments, Visual Communications Inc., Howard W. Sams & Company, and the Author.

PRINTED IN THE UNITED STATES OF AMERICA

9 8 7 6 5 4 3 2 1

Contents

This book is for my mother
and Prof. Ron Emery

About The Author

Vaughn D. Martin is a senior electrical engineer with the Department of the Air Force. Previously he worked at Magnavox and ITT Aerospace/Optics, where he acquired his fascination with optoelectronics. He has published numerous articles in trade, amateur radio, electronic hobbyist, troubleshooting and repair, and optoelectronics magazines. He has also written several books covering a wide range of topics in the field of electronics.

Foreword

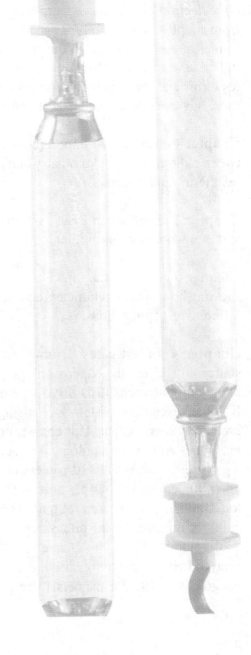

Foreword

Optoelectronics is an encyclopedic but practical coverage of optoelectronics. It does this with experiments, projects and many practical commercially available optoelectronic products. Collectively, this approach addresses readers attempting to educate themselves. Each chapter has quizzes which allow you to verify your progress and proceed at your own desired pace. There are thought-provoking quiz questions for which Appendix C provides the answers.

Optoelectronics, Volume 1 purposely progresses from simple to more complex aspects of optoelectronics, presenting you with the opportunity to familiarize yourself with the numerous optoelectronics terms and theories before progressing toward more advanced topics and experiments.

Chapter 1 makes the all important distinction between radiometry and photometry and defines optical terms and concepts.

Chapter 2 examines an LED's internal structure and semiconductor phenomena.

Chapter 3 covers testing optoelectronic devices, showing how to use photometers and lenses in nine different test applications.

Chapter 4 investigates driving visible light sources. It begins the "hands on" portion of the book with numerous LED drive schemes such as IC logic, discrete transistors and pulsing LEDs for greater apparent output. This chapter covers LED lens types and the psychological effects of color, e.g. the eye detects only 10 percent of red — the other 90 percent is invisible heat — thus the distorted illusion of apparent output. There are circuits for matching color intensities of single and dual colored LEDs. The chapter next examines arrayed LEDs, e.g. driving discrete and integrated bar graphs in different manners, three of which are:

1. All LEDs lighting as the voltage drives up the bar.
2. Just the top LED lighting.
3. You may flash the column with only the bottom LED on in an "exclamation point" attention getting mode.

There are circuits for greater measurement and display resolution by cascading LED bar graphs and their driver ICs and by "dithering," a light intensity modulating technique for duty cycle control. The 7-segment display drive circuits use the venerable 7447 TTL decoder/driver, and more modern CMOS ICs. There are techniques to overcome many of the annoying traits of some drivers. These include a lack of brightness control and the limited character set, preventing display of hexadecimal numerals. RAM and ROM decoders to drive dot matrix and "British flag" or 16-segment displays and 5 x 7 dot matrix displays conclude the circuit. Dot matrix displays allow you to customize your own character set. The only limiting factors are the resolution inherent in a 5 x 7 display's 35 dot format, and your imagination and creativity.

Chapter 5 covers basic theory and applications of photocells (light dependent resistors), photodiodes and photomultipliers.

Chapter 6 discusses CRT color systems and reviews established standards.

Chapter 7 examines:

1. Lenses and optical filtering to enhance front panel legends.
2. Nullifying the adverse effects of direct sunlight on displays.
3. Display contrast enhancement and 4) night vision goggles.

Chapter 8 introduces elementary optical math, including the Basic Lensmaker's Equation, and explains "focusing" light with a lens and various optical components such as prisms, mirrors and diffraction grating. The proper lens arrangement adds range to IR transmitters and sensitivity to IR receivers. Ideally, it can allow you to transmit data up to 150 feet.

The three appendices provide:

1. An extensive glossary of optoelectronics terms with which you may not be totally familiar.
2. A list of optoelectronic related products and suppliers.
3. Answers to all quiz questions.

Throughout the book, commercially available products help make the theory "come alive," bridging the gap between the "paper" theory of a text and more practical aspects of optoelectronics. Equations are minimally used, and only when no other means of explanation can clearly illustrate a point.

You might also note throughout the book this author's enthusiasm for the history of science and technology. If you share this interest in "The Dead Scientist's Society" (which, incidently, is an affectionate nickname for INVENTORLABS), you might be interested in a fascinating $39.95 3-D CD ROM available from Houghton Mifflin Interactive, Somerville, MA 617-351-3333. This particular CD ROM recreates, in the comfort of your home lab, the experiments of such great American inventors as Thomas Edison, Alexander Graham Bell, and James Watt. You can meet the scientists and examine their notes, sketches and scrapbooks, which were obtained from films, photographs and recordings. The experiments and the lab equipment are historically accurate and allow you to analyze their possible results.

Welcome to the field of Optoelectronics, an exciting technology which is useful, vitally important, rapidly emerging and constantly evolving.

Individuals from many optoelectronic companies helped me in my attempt to make this book useful and practical. I'll inevitably slight someone by their omission, but thanks to the following:

Max Bernard of Hewlett Packard supplied over 50 reference publications and considerable photos and camera ready artwork.

Jack Berlien of Texas Instruments supplied a complimentary TSL230 evaluation module.

Joe Howard and Peter Keller of Tektronix.

Karen Bosco of Motorola.

Robert Angelo of International Light.

Mike Reelitz of SUNX.

John Savage of Visual Communications Co.

Gary Baker of Dialight.

Yuval Tamari of Centronic.

Joseph Blandford of Miltec Corp.

Sue Casacia of ETI Inc.

Lynn O'Mara of Irvine Sensors.

Laurie Bass of Raytek.

Registered Trademarks and Tradenames

Windows™, Windows 3.1™, Windows for Work Groups™, Windows™ and Windows 95™ are registered trademarks of Microsoft.

LiteBug™ is a registered trademark of International Light.

Krylon™ is a registered trademark of Borden, Inc.

PAN X™ is a registered trademark of Kodak.

Cliplite™ and Cubelite™ are registered trademarks of Visual Communications Co.

PRISM CBI™ is a registered trademark of Dialight.

Chromafilter™ is a registered trademark of Panelgraphic Corp.

Plexiglas™ is a registered trademark of Rohm and Haas.

Light Control Film™ is a registered trademark of 3M Corp.

UVtron™ is a registered trademark of Hamamatsu.

Megga-Flash™ and Lite-Pac™ are registered trademarks of EG&G Electro-Optics.

UltraViolet Sensometer™ is a registered trademark of South Seas Trading Company.

CopperTone™ is a registered trademark of Schering-Plough.

UVICURE Plus™, SpotCure™ and UVIMAP™ are registered trademarks of ETI.

SharpEye 20/20UB™ is a registered trademark of Spectrex.

Microtops™ and UV-Biometer™ are registered trademarks of Solar Light.

TouchTone™ is a registered trademark of AT&T.

Sunverter™ is a registered trademark of Abacus.

TekLumaColor™ is a registered trademark of Tektronix.

Light Control Film™ is a registered trademark of 3M.

PRISM CBI™ and Optopipe™ are registered trademarks of Dialight.

Lexan™ is a registered trademark of General Electric.

Ultralume™ is a registered trademark of Philips.

Silver Saver™ is a registered trademark of The Orchid Corp.Newton™ is a registered trademark of Apple.

Transit^{tm} and TranXit Pro^{tm} are registered trademarks of Puma Technologies.

LaserJet^{tm} is a registered trademark of Hewlett-Packard.

Thinkpad 755^{tm} is a registered trademark of IBM.

THERMO-DUCER^{tm}, Thermalert^{tm} and Raytek Field Calibration and Diagnostic Software^{tm} are registered trademarks of Raytek Corp.

A-690 Plus^{tm} is a registered trademark of UVP, Inc.

Optoelectronics

6

Chapter 1

Optoelectronics Terminology and Concepts

Chapter 1
Optoelectronics Terminology and Concepts

Electromagnetic radiation covers a wide spectrum of both visible (photometric) and non-visible (radiometric) light. A light bundle consists of trillions of photons at different wavelengths, times and frequencies. Despite this true but ostensibly chaotic description, light is quite predictable, measurable and useful.

The Physical Basis of Light

Light is produced by the energy released when atoms are excited by heat, chemical reaction, or other means. Light is both a wave and a particle, and travels through space as electromagnetic waves. Due to this wave motion, each "color" has a unique wavelength. This wavelength is the distance a wave travels in one cycle. Since light wavelengths are so short, we express them in nanometers. (See *Table 1-1*) For example, the color green, the most visible light, has an approximate wavelength of 565 *nm*. Although it is a wave, light has neither mass nor charge. A light particle, called a photon, is classified as a particle because it is the smallest "piece" of light. This is analogous to an electron being the smallest negatively-charged "piece" of an atom.

The Inherent Problems of Optoelectronic Terms

Before we start investigating optoelectronics, we must familiarize ourselves with this technology's terms. Optoelectronic terms are admittedly far more numerous and complex than the common parameters and unit symbols associated with electronics, such as the watt, volt, ohm, etc. This is because optoelectronics combines two primary technologies: optics and electronics. *Table 1-2* shows modern terms versus their equivalent, outdated terms. Confusion exists due to the continued misuse of these terms and the numerous ways in which these unit symbols can be stated. There are many more

ELECTROMAGNETIC SPECTRUM

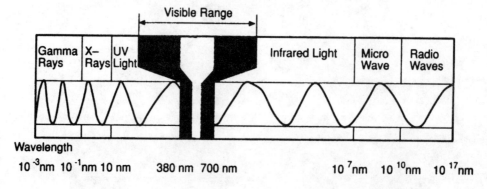

Table 1-1. *Light wavelengths expressed in nanometers.*

Term, Definition	Symbol	Defining Equation	New Term (Old Term)	Symbol	SI Units, Abbr.	New Term (Old Term)	Symbol	SI Units, Abbr.	Equivalent
	GENERIC		RADIOMETRIC			PHOTOMETRIC			
FLUX, rate of flow of energy, Q. Q_e – radiant, Q_v – luminous, Q_q – photon [see note 1]	ϕ	$\dfrac{dQ}{dt}$	Radiant Flux (Radiant Power)	ϕ_e (P)	watts, W	Luminous Flux (Luminous Flux)	ϕ_v	lumens, lm	
INCIDANCE, flux per unit area on a reception surface	E	$\dfrac{d\phi}{dA}$	Radiant Incidance (Irradiance)	E_e (H)	watts per sq. meter, W/m^2	Luminous Incidance (Illuminance or Illumination)	E_v (E)	lux, lx [see note 3]	lumens per sq. meter, lm/m^2
EXITANCE, flux per unit area from an emitting surface	M	$\dfrac{d\phi}{dA}$	Radiant Exitance (Emittance)	M_e (W)	watts per sq. meter, W/m^2	Luminous Exitance	M_v	lumens per sq. meter, lm/m^2	[see note 2]
INTENSITY, flux per unit solid angle from a remote source	I	$\dfrac{d\phi}{d\omega}$	Radiant Intensity (Radiant Intensity)	I_e (J)	watts per steradian, W/sr	Luminous Intensity (Luminous Intensity)	I_v (I)	candelas, cd	lumens per steradian, lm/sr
STERANCE, flux per unit solid angle per unit area of emitting surface at angle θ with respect to surface normal.	L	$\dfrac{dI}{dA \cos\theta}$ $\dfrac{d^2\phi}{d\omega(dA\cos\theta)}$	Radiant Sterance (Radiance)	L_e (N)	watts per steradian per sq. meter, $W/sr/m^2$	(Luminous Sterance) (Luminance)	L_v (B)	candelas per sq. meter, cd/m^2 [see note 3]	lumens per steradian per sq. meter, $lm/sr/m^2$ nit, nt

Note 1. Quantametric terms use the prefix word "photon" and their symbols have a subscript, "q".
Note 2. Lux and other units of luminous incidance DO NOT APPLY to luminous exitance.
Note 3. Other units in Table 7.2-2.

Table 1-2. *Terms, definitions, and explanations of both newer and older optoelectronic energy unit symbols. (Courtesy of Hewlett-Packard)*

ways of stating terms than just the English versus the metric system; but these terms aren't impossible to learn, so let's tackle them.

The Conventional Terms of Optoelectronics

Photometry refers to visible light and its characteristics. Photometry deals with flux (energy/time), in lumens, at visible wavelengths. *Radiometry* deals with flux, in watts, at non-visible wavelengths, typically IR (infrared) and UV (ultraviolet). Except for the difference in units of flux, both of these units have identical geometric concepts. The radiant prefix describes radiometric units. The luminous prefix describes photometric units. As we will later discover, since photometry deals with flux, in lumens, at wavelengths that are visible, its unit symbols have a subscript "v," which stands for *visible*. Radiometric unit symbols have a subscript "e," which stands for *radiant energy*.

A History of Light and Photoelectrics

In 1666, Sir Isaac Newton proposed that light was composed of tiny "corpuscles" (particles) and traveled in a straight line. Simultaneously, Dutch physicist Christian Huygens argued that light traveled in waves. In 1860, the British physicist James Clerk Maxwell proposed that light traveled in electromagnetic waves, not mechanical waves as sound does. In 1900, Max Planck characterized light as composed of small packets of energy called *quanta*, and was corpuscular in nature. Therefore, both Newton and Huygens were correct.

The first significant observations of radiation and electricity were made by Gustav Hertz in 1887. Hertz observed that under exposure to radiant light, certain surfaces liberated electrons. In 1900, Max Planck further proposed the revolutionary theory time that light contained energy in discrete bundles or *quanta*, and the quanta were composed of photons. In 1905, Einstein formulated that the energy content of photons was directly proportional to the frequency of the light. Each photon has energy and a frequency, described by the following equation:

$$E = h \cdot f$$

E = energy in Joules, h = Planck's constant of $6.63 (10^{-34})$, and f = frequency expressed in Hertz.

CHARACTERIZING VISIBLE LIGHT

The colors red, orange, yellow, green, blue, indigo, and violet represent the light spectrum visible to humans. It is particularly noteworthy that green and yellow are in the middle of this spectrum, and are therefore the most visible colors. This is what has obviously prompted fire engine manufacturers to abandon the practice of painting these vehicles the traditional "fire engine" red — many engines are now painted yellowish-green.

Now that we understand the basis for optoelectronics, let's characterize light parameters associated with both photometric (visible) and radiometric (non-visible) light indicators — predominately LEDs. A Hewlett-Packard LED photometry application note jokingly equates "evaluation of beauty contestants" to judging an LED's light output. This is an accurate description; nonetheless, there are established scientific means by which to judge this illusive subject. First, let's briefly discuss human light perception.

The Problem Humans Inherit

The inadequacy of attempting to quantify any physically-perceived phenomenon stems from a person's two nonlinear or logarithmic sensors. These are our eyes and ears. The logarithmic dB or decibel units measure sound intensity. Light is also logarithmic. In low ambient-lighted settings, a small LED spot saturates the eye's retinal response. The human eye also detects certain frequencies (colors) much better than others. Together, these factors confound the problem of measuring/characterizing intensity and other factors associated with visible light sources.

Realizing these human shortcomings, it may seem a moot point to scientifically characterize LEDs. This is especially true after a sophisticated instrument characterizes an LED; and the real "acid test" is still ultimately in how it appears to your eye. Sophisticated instruments nonetheless still aid in

the manufacturing process, if for no other reason than to ensure the uniformity of optical parameters. Later in this book is an experiment with green and red LEDs which will graphically demonstrate this.

Setting the Ground Rules

To adequately measure light parameters, we must agree upon the definitions of certain terms, apparatus setups, and criteria for the selection of equipment. This may well appear obvious; however, there is a lack of uniformity within the industry for LED testing and rating.

DEFINITIONS OF TERMS

This is a very complex and confusing subject, since improperly using certain terms partially contributes to many problems in optoelectronics. Modern terms have replaced many of the older terms, such as footcandle. *Table 1-2* presents both older and newer terms. This enables you to understand older literature and optoelectronic component data sheets that are still being circulated.

As previously stated, photometry deals with flux (in lumens) at wavelengths that are visible; its unit symbols have a subscript "v," which stands for visible. Radiometric unit symbols have a subscript "e," which stands for radiant energy. Radiometry deals with flux (in watts) at all wavelengths of radiant energy, including infrared LEDs and UV light. Another way to distinguish the two is that photometry unit symbols have the prefix "luminous," while radiometry unit symbols have the prefix "radiant." *Flux* describes the rate at which energy passes to, from, or through a surface or other geometrical entity. *Figure 1-1* and *Table 1-2* describe optoelectronic generic terms and symbols. Flux primarily relates to other terms. Few applications use all of the flux available from a source. The same is true for exitance — this term describes the flux per unit area leaving a source of finite area. The only practical situation in which this is not true is with a receptor so tightly coupled to a source that

In general, confusion is averted by use of self-explanatory units, such as: lumens per square meter for luminous incidance; and, candelas per square meter for luminous sterance. At times, for brevity, where the risk of confusion is negligible, other units used are:

Luminous Sterance		Lambert L	Footlambert fL	Apostlib asb	*stilb sb	cd/ft^2	**cd/m^2	nit
Unit, Abbr.	Equivalent							
Lambert, L	$1/\pi$ cd/cm^2 =	1	929	10,000	.3183	295.7	3183	3183
Footlambert, fL	$1/\pi$ cd/ft^2	.001076	1	10.76	.0003426	.3183	3.426	3.426
Apostilb, asb	$1/\pi$ cd/m^2 =	.0001	.0929	1	.00003183	.02957	.3183	.3183
*Stilb, sb	cd/cm^2 =	3.1416	2919	31,416	1	929	10,000	10,000
Candelas per sq. foot	cd/ft^2 =	.003382	3.1416	33.82	.001076	1	10.76	10.76
**Candelas/sq. meter	cd/m^2 =	.00031416	.2919	3.1416	.0001	.0929	1	1

Luminous Incidance		*Phot ph	*Footcandle fc	**, *Lux lx
Unit, Abbr.	Equivalent			
*Phot, ph	lm/cm^2	1	929	10,000
*Footcandle, fc	lm/ft^2	.001076	1	10.76
**, *Lux, lx	lm/m^2	.0001	.0929	1

*CIE Unit
**Recommended SI Unit
Many of the other units
are in common usage, but
efforts are being made to
standardize on the SI units.

USE OF TABLE: In any row, the quantities are all equal; for example, in the table for luminous sterance, cd/m^2 = .00031416 L = .2919 fL = 3.1416 asb = .0001 sb = .0929 cd/ft^2 = 1 nt; in the table for luminous incidance, lx = .0001 ph = .0929 fc.

Figure 1-1. *Generic optoelectronic terms and unit symbols.*
(Courtesy of Hewlett-Packard)

virtually all of the flux leaving the source enters the receptor. This rare condition does occur in sandwich-type optoisolators. More on this in a later chapter.

The Relationship Between Radiant and Luminous Flux

What is the relationship between luminous and radiant flux? The answer is the *luminosity function* (V), also called the *standard observer curve,* or *CIE curve.* CIE is the Commission Internationale de l'Eclair, a French commission which is the equivalent of America's NIST, formerly called the National Bureau of Standards. *Figure 1-2* shows the CIE curve. Note that at the peak wavelength of 555 nanometers (5,550 Ångstroms) the conversion factor is 680 lumens = 1 watt. Incidentally, also note that 1 nanometer (nm) equals 10 Ångstroms (Å).

A Practical Example of Using the CIE Curve

Two common LEDs, a red and a green one, demonstrate the curve's usefulness. Again, the CIE curve shows how your eye perceives or detects 555 nm or green most easily. A green LED with less current flowing through it and less light emitted can appear brighter than a red LED with ten times greater radiant energy.

Let's use a red LED and a green LED from the same company. Refer again to the CIE curve in *Figure 1-2*. Now, let's perform the calculations. The red LED emits about 180 mW of radiant energy at 660 nm or 41.4 lumens/watt. Dividing this yields 7.45 mLumens. The green LED produces only 10 mW of radiant energy at 570 nm. This frequency corresponds to 649 Lumens per watt. Therefore, multiplying these together yields 6.49 mLumens. Though the red LED emitted 18 times as much radiant energy, the eye virtually perceives the two

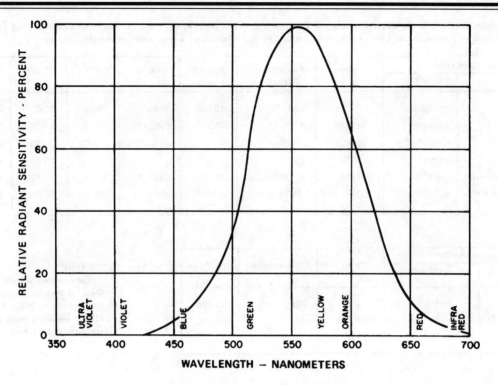

Figure 1-2. *The CIE curve.*

(7.45 vs. 6.49 mLumens) as having equal brightness. The eye tends to magnify the green to appear as bright as the red LED. The human eye responds to only about 10% of the energy in the color red. The remaining energy is lost as heat toward the infrared end of the light spectrum.

Three Highly-Practical Optoelectronic Terms

There are only five units for radiant energy. (See *Table 1-2* again.) There are, however, three more terms which will help your understanding of practical electro-optical measurements. These are *incidance*, *intensity*, and *sterance*.

Incidance

This has the same unit symbols as *exitance*, but is vastly different since it ignores the direction taken by exiting flux. It is therefore most useful in describing photodetector properties. There are certain constraints on measuring the intensity of an LED. If the measured source has a plane of emission with respect to density which you can define, the problem's solution is simple. You merely need

a *photometer* — an instrument which measures visible light quantity (flux) and brightness (intensity), to read the incidance (E_v) at a distance (d) from the source.

Intensity

This is extremely useful in both photometric and radiometric applications; in photometric applications, we use the candela unit symbol. Since flux traveling through space is usually divergent, you can usually define an equivalent point from which it diverges in terms of a solid angle and the intensity of that point. This equation describes luminous intensity:

$$I_v = E_v \, d^2$$

This is valid only if d is large enough that this equation —

$$\omega = A/d^2$$

— describes a solid angle. If we want the error to be 1% or less, the distance (d) must be at least ten

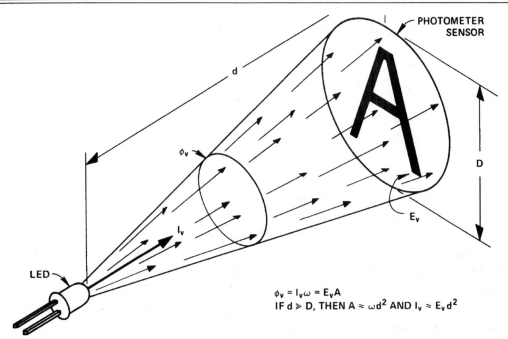

$$\phi_v = I_v\omega = E_vA$$
IF $d \gg D$, THEN $A \approx \omega d^2$ AND $I_v \approx E_v d^2$

(a) RELATIONSHIP BETWEEN LUMINOUS INTENSITY, I_v, OF AN LED
AND LUMINOUS INCIDANCE, E_v, AT THE PHOTOMETER

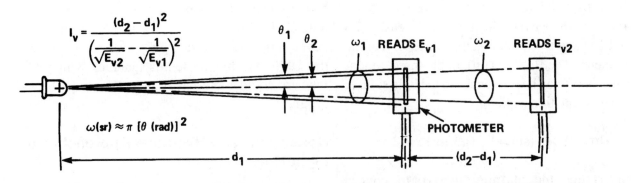

$$I_v = \frac{(d_2 - d_1)^2}{\left(\dfrac{1}{\sqrt{E_{v2}}} - \dfrac{1}{\sqrt{E_{v1}}}\right)^2}$$

$$\omega(sr) \approx \pi \left[\theta\ (rad)\right]^2$$

(b) APPROXIMATE TWO-POINT METHOD ASSUMES $\varphi_1/\omega_1 = \varphi_2/\omega_2$

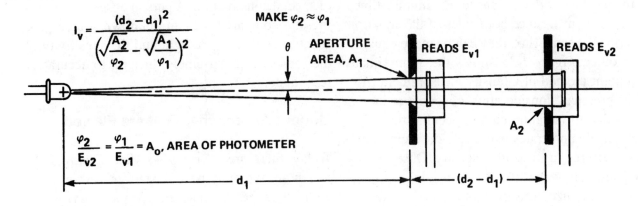

$$I_v = \frac{(d_2 - d_1)^2}{\left(\sqrt{\dfrac{A_2}{\varphi_2}} - \sqrt{\dfrac{A_1}{\varphi_1}}\right)^2}$$

MAKE $\varphi_2 \approx \varphi_1$

$$\frac{\varphi_2}{E_{v2}} = \frac{\varphi_1}{E_{v1}} = A_o,\ \text{AREA OF PHOTOMETER}$$

(c) PRECISE TWO-POINT METHOD USES TWO APERTURES, A_1 AND A_2
SO THE SOLID ANGLE IS NEARLY THE SAME AT EACH POINT.
THIS REDUCES ERROR DUE TO RADIATION PATTERN PECULIARITIES.

Figure 1-3. *Making intensity measurements with a photometer. (Courtesy of Hewlett-Packard)*

times the diameter of the source. This is called the *ten diameters rule.*

Intensity Measurements

When you do not define the distance, you may make intensity measurements at two different distances. With this "approximate" method, making theta (Θ) small improves accuracy, up to a point. (See *Figure 1-3b.*) This point is when Θ becomes so small the photometer does not receive enough flux, which makes the S/N (signal-to-noise) ratio inadequate for an accurate reading. This is analogous to making sensitivity measurements on a receiver at such low levels that the magnitude of the noise overcomes or "swamps" the signal.

If the light source has a plane of emission from which you can define intensity, all that you need to do is read the incidance (E_V) and know the distance (d) from the light source. *Figure 1-3a* shows the relationship between the luminous intensity (I_V) of an LED and the luminous incidance (E_V) at the photometer. The flux, φ_V (lm), in the three dimensional "cone" may be referenced to the light source, which this equation explains:

$$I_V \text{ (lm/sr) x } \omega \text{ (sr) = } E_V \text{ (lm/m}^2\text{) x } A \text{ (m}^2\text{)}$$

I_V (lm/sr) times the solid angle ω (sr) of the cone is the intensity, or you may refer it to the light sensor as the product of the incidance, E_V (lm/m²), times the area, A (m²). Since both sides of the equation describe the same flux, they can be made equal to each other. *Figure 1-3b* shows the approximate two-point method of intensity measurement. Another practical consideration concerns reducing errors due to radiation pattern peculiarities. *Figure 1-3c* demonstrates using two different diameter apertures to measure intensity. Observe this setup and follow the three-step procedure outlined here to minimize measurement errors.

The Three-Step Intensity Measurement Procedure

1. Refer to *Figure 1-3c* and place the photometer at a distance greater than 100 times the diameter of a **SMALL** bright light source, such as an LED.
2. Place A_1 over the photometer while taking a reading, which you'll call E_1. Then A_1 is equal to A_0 (E_1/E_0).
3. Now place A_2 over the photometer; this reading is E_2. Equate A_2 equal to A_0 times (E_2/E_0). If you make several observations at different distances, it will improve the accuracy of this procedure.

Sterance

Luminous sterance is the basis for distinguishing an object from its background color — this is a popular measure of contrast and photographic exposures. As an example, you can use it to see gravy spots on a green tie despite the spot having the same luminous sterance as the tie. If the same spot was on a brown tie, the luminous sterance would have to be greater than a ratio of 2:1 to be visibly noticeable. An even more practical example is the development of a special phosphorescent ball for the 1996 Wimbledon tennis matches which TV viewers could more easily detect from its background.

Constraints and Historical Approaches to Light Measurement

The original standard for candlepower was a candle of special construction. A person observed it burning and compared it to an unknown source. The spectral quality of the light and the sensitivity of the sensor were extremely difficult to accurately reproduce.

Modern Approaches Still are Flawed

Today, there are *thermopiles* (See *Figure 1-4*), which are typically heated, stacked carbon discs with filters that simulate the eye's sensitivity. Despite the improvements, light measurement is still difficult because of stray light effects, color temperature shifts, nonuniform light distribution, the drifting of sensors and other variables. The National Bureau of Standards (now NIST) issued a

Figure 1-4. A typical carbon stacked thermopile.

paper stating that, at best, uncertainties of 4.4 to 4.7% exist when attempting to measure light output from a solid-state lamp, such as an LED. Therefore, if your measurements come within 1% of someone else's measurements, do not gloat with pride — you both could be wrong. However, if your results deviate by more than 10%, then something is definitely wrong with your procedure, or with the LED being tested!

The Concept of Color Temperature

In all photometric operations, it is necessary for the lamps to be at a known color temperature. Tungsten lamps have spectral characteristics defined in color temperature. This is the equivalent color of heating a perfect reflector and absorber of radiant energy, called a *black body* (See *Figure 1-5*) — a radiometer in front of a black body, which isn't black at all. Later in the book we will build two color temperature meter projects — devices which have a continuous spectral output. For normal work, use standard Tungsten lamps which run at a color temperature of 2854 K. The letter *K* stands for Kelvin, which is a temperature scale starting at the theoretical temperature of absolute zero, or - 273.15°

C. This is zero K: note the absence of a degree sign, which is the correct method! Therefore, 2854 K is about 2854 - 273 = 2581°C, or the equivalent color if you were to heat the black body to that temperature.

Special Lab Lamps

You will mostly use Tungsten lamps as secondary standards, which have been carefully measured for candlepower output at a carefully controlled voltage and current. Primary standards exist only at NIST (National Institute of Standards and Technology), formerly called the NBS or National Bureau of Standards. These are the ultimate in accuracy; naturally, no industries or individuals readily have access to them. The secondary standards, though, are available from NIST and several private testing laboratories. Their output is in HCP (horizontal candlepower), which means you measure the output in a normal plane perpendicular to the lamp's vertical axis and at a height in line with the filament.

Usually, these special lamps have front and back marks for orientation, which helps you obtain re-

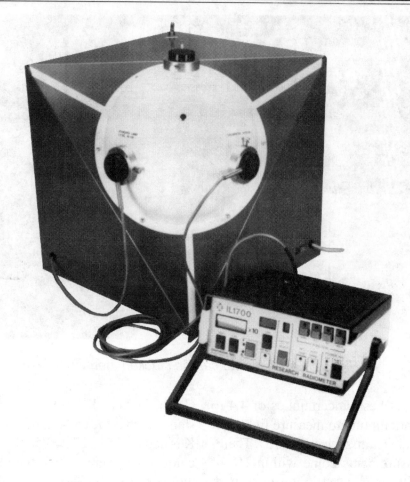

Figure 1-5. *A radiometer and a blackbody.
(Courtesy of International Light)*

peatable results. Following this convention, you can calibrate and reproduce a variety of illumination levels by varying the distance between the photocell and the lamp. The following equation calculates footcandles. Note that the light intensity is inversely proportional to distance, just as we logarithmically perceive sound intensity:

$$HCP/D^2 = Footcandles$$

Using Foot-Lamberts

There are many times when candle measurements are not appropriate and brightness measurements are. These are cases in which light reflects from a secondary surface, causing it to no longer be a point source but rather an area source. The common unit to express brightness is the *foot-Lambert*. A foot-

Lambert is the brightness of a uniformly diffusing surface reflecting 100% of the light falling upon it when illuminated with one footcandle. Therefore, if one footcandle illuminated a surface which had a reflectance of 80%, the brightness of the surface would be 0.8 foot-Lamberts. A common method of simulating brightness sources is by placing an optical filter (a sheet of opal glass) between the photocell and lamp. The glass is already calibrated for a conversion factor of foot-Lamberts on one side of the glass, and of footcandles of illuminating power on the opposite side.

Photometry Terms

A *lumen* (lm) is the basic unit of photometry. It describes the luminous flux radiated from visible objects and sources. Since lumens only describe the

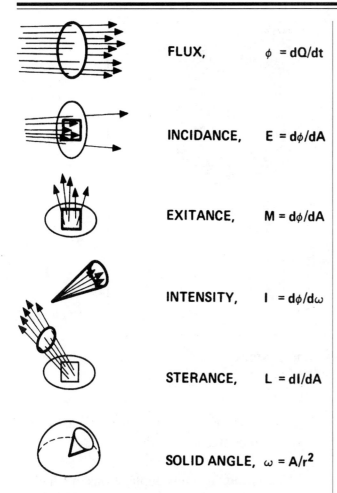

FLUX,	$\phi = dQ/dt$	
INCIDANCE,	$E = d\phi/dA$	
EXITANCE,	$M = d\phi/dA$	
INTENSITY,	$I = d\phi/d\omega$	
STERANCE,	$L = dI/dA$	
SOLID ANGLE,	$\omega = A/r^2$	

Figure 1-6. Defining solid geometry-related optoelectronic terms.
(Courtesy of Hewlett-Packard)

total emitted luminous flux, with no regard to the flux direction with respect to the viewer, they are inadequate to totally describe a light source. For LEDs in particular, it is more useful to consider the flux-per-unit solid angle. You may recall from geometry that there are two π or 6.28 radians in a circle. This equates to approximately 57.296° in a radian. This, though, is naturally on a flat surface or a plane, whereas you obviously reside in a three-dimensional world. Therefore, a solid geometry (three-dimensional) steradian (resembling a 57.3°-shaped ice cream cone) is more appropriate. One lumen per steradian has a luminous intensity (I_v) of one candela (cd). (See *Figure 1-6.*)

In LED characterization, the two terms you most commonly use are *foot-Lamberts* and *millicandelas*. The millicandela (mcd) is one thousandths of a candela. The foot-Lambert (fL) describes intensity per unit of area. Intensity per unit area is *luminance* (L_v). It would be just as useful to give luminance in candelas per square centimeter; however, common usage popularized the foot-Lambert. Also, π foot-Lamberts equals one candela per square foot. The following two equations help explain this better:

Units Expressed in the English System
$$I_{v/\pi} = 1\ L_v A$$

I_v = Luminous intensity of sources in candelas, L_v = Luminance of source in foot-Lamberts (fL), and A = Area of source in square feet (ft^2).

Units Expressed in the Metric System
$$I_v = L_v A$$

I_v = same as previous equation, L_v = luminance of source in candelas per square meter (cd/m^2), A = Area of source in square meters (m^2).

Luminous incidance (E_v), formerly called *illuminance*, is a subtle LED characterization describing the luminous flux per unit area incident upon a surface from a point source. Mathematically, it is:

$$E_v = I_v/d^2$$

E_v = luminous incidance at surface in lumens per square meter (lm/m^2), I_v = luminous intensity of the source in candelas (cd), and d = source-to-surface distance in meters (m).

In the previous equation, if the distance (d) is in feet, the incidance is in lumens per square foot. One lumen per square foot is a footcandle (fc). Our previous tennis ball example showed how luminance makes an object visible by distinguishing it from its background. An LED background and filtering make it only necessary to have adequate luminance. In LED characterization, four factors make it ap-

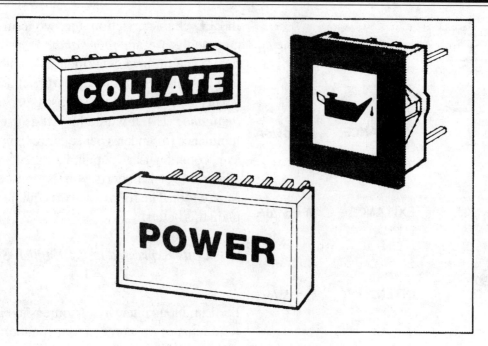

Figure 1-7. *Backlighting examples.*
(Courtesy of Hewlett-Packard)

parent why intensity (millicandelas) is superior to luminance (foot-Lamberts):

1. *Figure of Merit*. Due to manufacturing variances, there are apparent discrepancies in size resulting from variable degrees of diffusion, magnification, and luminance. Accordingly, the product of luminance times area is their figure of merit.

2. *Viewing Angle*. An LED's luminance is the same magnitude at any angle of view. What changes most, as the viewing angle varies, is the apparent area, expressing only minor changes in luminance.

3. *Measurement*. Luminance measurement requires sophisticated equipment to magnify, focus, and field-stop an image of the LED so that the photometric detector (an instrument which measures both brightness and intensity) receives flux only from an incremental portion of the LED's emitting surface. *Channel baffling*, though, does help this. Conversely, luminous intensity is very consistent and easily measured

with a calibrated detector and a distance scale. We shall do this later in the book.

4. *Illumination*. In a few applications in which LEDs illuminate a legend such as might appear on a front panel annunciator (See *Figure 1-7*), the level of luminance incidance that the LED can produce is the measure of its performance. Luminous incidence varies jointly as the luminous intensity of the source; this is true even when the inverse square law for distance does not apply. *Contrast* is related property. (See *Figure 1-7* again.) The following equation defines contrast:

**Contrast = The legend's light
+ background light/background light**

Measuring Photometry Parameters

Using the following equation, it is only necessary to know the distance from the LED to the calibrated photosensor. Then, we can obtain the luminous incidence (E_v). If this distance is large enough, the inverse square law applies, (See *Figure 1-8*,

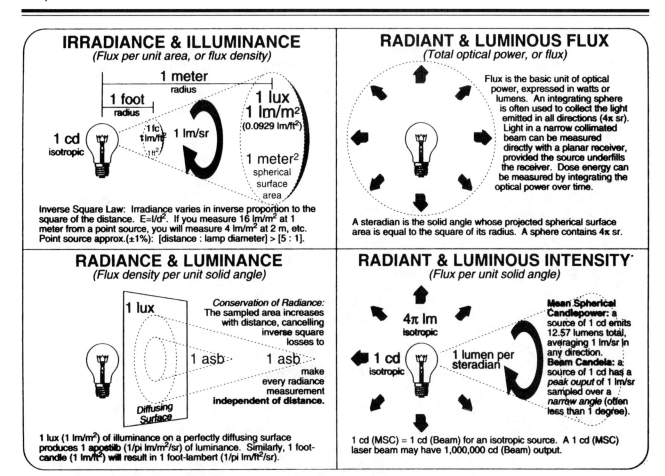

Figure 1-8. The inverse square law and optical concepts.
(Courtesy of International Light)

the upper left illustration), and the following equation states:

$$I_v = E_v \times d^2 \quad I_v = \text{luminous intensity of source in candelas (cd)}$$

Units in the English System
E_v = luminance incidence at
sensor in footcandles (fc).
(A *footcandle* is an old optoelectronics unit.)
d = distance-to-sensor in feet (ft)

Units in the Metric System
E_v = lumens per square meter (lm/m²)
d = distance-to-sensor in meters (m)

In Summary

When reviewing the various optoelectronic terms,

remember what they are describing. The following four definitions will help:

Flux = Energy/Time
Incidance and Exitance = Flux/Area
Intensity = Flux/Steradian(s)
Sterance = Intensity/Area x cosine of the light's angle of arrival of your viewing angle.

See *Figure 1-8* again, which illustrates many of these terms and concepts. *Table 1-3* shows how to convert from the newer accepted unit symbols to other units of measure.

POWER: *(Flux)*
1 watt (W)
    = 0.27 lm at 400 nm
    = 25.9 lm at 450 nm
    = 220.0 lm at 500 nm
    = 679.0 lm at 550 nm
    = 683.0 lm at 555 nm
    = 430.0 lm at 600 nm
    = 73.0 lm at 650 nm
    = 2.78 lm at 700 nm
1 lumen (lm)
    = 1.464×10^{-3} W at 555 nm
    = 7.958×10^{-2} candela (4π sr)
1 joule (J)
    = 1 watt•second
    = 10^{7} erg
    = 0.2388 gram•calories
1 lm•s
    = 1 talbot (T)
    = 1.464×10^{-3} joules at 555 nm

IRRADIANCE: *(Flux / Area)*
1 W/cm²
    = 10^{4} watts/m²
    = 6.83×10^{6} lux at 555 nm
    = 14.33 g•calories/cm²/min
1 lm/m²
    = 1 lux (lx)
    = 10^{-4} lm/cm²
    = 10^{-4} phot (ph)
    = 9.290×10^{-2} lm/ft²
    = 9.290×10^{-2} foot-candles (fc)

INTENSITY: *(Flux / steradian)*
1 W/sr
    = 12.566 watts (isotropic)
    = 683 candela at 555 nm
1 lm/sr
    = 1 candela (cd)
    = 12.566 lumens (isotropic)
    = 1.464×10^{-3} watts/sr at 555 nm

RADIANCE: *(Flux / Area / sr)*
1 W/cm²/sr
    = 6.83×10^{6} lm/m²/sr at 555 nm
    = 683 cd/cm² at 555 nm1
lm/m²/sr
    = 1 candela/m² (cd/m²)
    = 1 nit
    = 10^{-4} lm/cm²/sr
    = 10^{-4} cd/cm²
    = 10^{-4} stilb (sb)
    = 9.290×10^{-2} cd/ft²
    = 9.290×10^{-2} lm/ft²/sr
    = 3.142 apostilbs (asb)
    = 3.142×10^{-4} lamberts (L)
    = 2.919×10^{-1} foot-lamberts (fL)

Table 1-3. A table for converting from one optical unit symbol to others.
(Courtesy of International Light)

Chapter 1 Quiz

1. Photometry deals with visible light, T or F.
2. Radiometry deals with non-visible light, T or F.
3. The luminous prefix describes radiometric units, T or F.
4. The radiant prefix describes photometric units, T or F.
5. When you do not define distance, you can make intensity measurements at two different distances, T or F.
6. Part of the problem with accurately judging visible colors and light results from:
    A. Our eyes are logarithmic sensors.
    B. In low light settings, an LED spot saturates our retinal response.
    C. Many of us have distinct color preferences.
    D. Both A and B.
7. Which is an attempt to characterize how our eyes perceive color?
    A. The CIE curve.
    B. Color blindness tests.
    C. Colored optical filters.
    D. All of the above.
8. The human eye tends to magnify which color?
    A. Red.
    B. Orange.
    C. Blue.
    D. Green.
9. In judging photometric parameters what type lamps do we use?
    A. Halogen.
    B. Laser.
    C. Tungsten.
    D. Neon.
10. When light reflects from a secondary surface, it is no longer a point source, but rather a:
    A. Diffused source.
    B. Area source.
    C. Dark source.
    D. None of the above.
11. What is the basic unit of photometry?
    A. A Watt.
    B. A lumen.
    C. A steradian.
    D. A radian.
12. In characterizing LED parameters, the two terms we most commonly use are:
    A. Foot-Lambert and millicandela.
    B. Foot-Lambert and flux density.
    C. Foot-Lambert and steradian.
    D. Viewing angle and flux.
13. Luminance times area is:
    A. The figure of merit.

B. The viewing angle.
C. The illumination factor.
D. None of the above.

14. The "ten diameters" rule ensures:
    A. Against errors greater than 10% due to distance.
    B. Against errors from different LED lens types.
    C. Against errors from changing measurement systems.
    D. Against errors from changing photosensor types.

15. Who first proposed that light was composed of tiny particles he called "corpuscles?"
    A. Einstein.
    B. Huygens.
    C. Planck.
    D. None of the above.

16. Who first proposed that light travels in waves?
    A. Einstein.
    B. Huygens.
    C. Planck.
    D. Hertz.

17. Who first observed that certain surfaces liberate electrons when exposed to light?
    A. Hertz.
    B. Huygens.
    C. Planck.
    D. None of the above.

18. Who first proposed that light travels in electromagnetic waves?
    A. Maxwell.
    B. Huygens.
    C. Planck.
    D. Hertz.

19. Who first proposed that light contains discrete bundles of energy called quanta?
    A. Einstein.
    B. Huygens.
    C. Planck.
    D. Hertz.

20. Who first proposed that light's energy content is directly proportional to its frequency?
    A. Einstein.
    B. Huygens.
    C. Planck.
    D. Hertz.

Chapter 2

Optoelectronics Semiconductor Phenomena

Chapter 2
Optoelectronics
Semiconductor Phenomena

This chapter stresses practical semiconductor phenomena and trade-offs in LED manufacturing techniques. Optoelectronic semiconductor phenomena dominantly influence LED device fabrication. For example, diffracted light traveling through different media within an LED influences the material, shape, and dimensions of the LED's plastic dome encapsulation. The dome influences photometric efficiency. Using silicon (Si) as the dopant in the LED's P and N regions shifts the wavelength several hundred Ångstroms. Less readily, the dome absorbs this longer wavelength, increasing photometric efficiency (light output).

Any diode, regardless of its type, has P-type and N-type adjacent semiconductor materials. The N-type material has an excess of electrons. The P-type material has an excess of holes. A free electron's orbit is larger and has more energy than a hole's orbit. After free electrons cross the P-N junction, they radiate energy and heat, and emit light in the process. Why don't germanium or silicon diodes emit light? *Figure 2-1* shows a diode with an opaque substrate, and an LED with a semitransparent substrate, one of three types of LED substrate materials:

1. Gallium-Arsenide (GaAs). Emits infrared (invisible) radiation.
2. Gallium-Arsenide Phosphide (GaAsP). Emits red or yellow light.
3. Gallium Phosphide (GaP). Emits red or green light.

An Overview of Optoelectronics History

When Bell Lab's William Schottky invented the transistor in 1949, he noticed that the light entering the case and striking the transistor's crystal structure generated a photoconductive charge. This increased the illuminated material's conductivity, causing erratic, unexplainable performance. He named the "transistor" from his further observance that it **trans**ferred resistance of a re**sistor** within a crystal. Today, we take advantage of this phenomenon and exploit it, creating the technology of *optoelectronics*.

Historically, optoelectronic emitter development has lagged behind optoelectronic sensor development by about 20 years, because visible light emission does not occur in silicon and germanium. Metallurgists and chemists have since developed compound materials, such as gallium phosphide (GaP), and the more complex gallium-arsenide phosphide (GaAsP). These materials emit the proper wavelength of visible light, which helps to create optoelectronic emitters.

The Light Mechanism

According to the universally-accepted Bohr model, an atom's nucleus has orbiting electrons. The number of electrons and protons determines the atom's electrical and chemical properties. These orbits, called *shells*, are arranged in an orderly fashion.

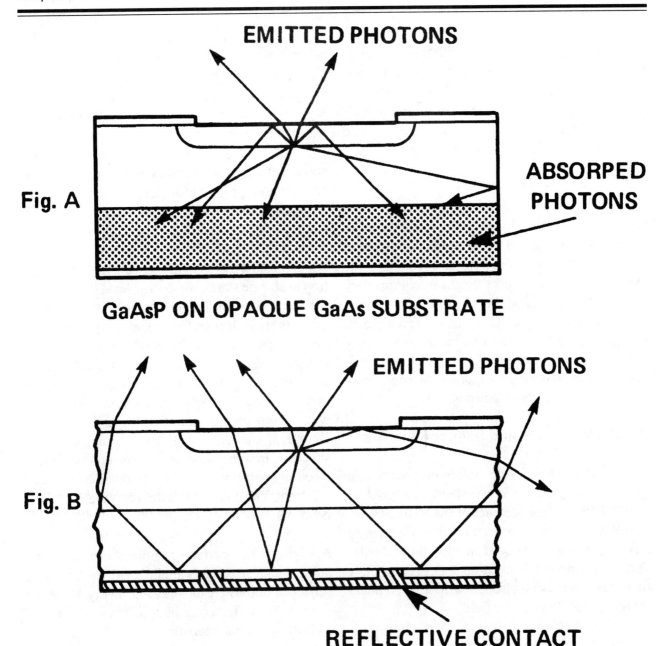

EMITTED PHOTONS

Fig. A

ABSORBED PHOTONS

GaAsP ON OPAQUE GaAs SUBSTRATE

EMITTED PHOTONS

Fig. B

REFLECTIVE CONTACT

GaAsP OR GaP ON TRANSPARENT GaP SUBSTRATE

Figure 2-1. *Photons escaping from an opaque and a transparent substrate.*
(Courtesy of Hewlett-Packard)

The innermost shell is K, which has two electrons. The L shell has eight electrons, and so on. As the shells progress outward from the nucleus, their names progress in the alphabet, starting with L as the first letter in the sequence.

The outer or *valence* shell has a maximum of eight electrons. All elements have outer or valence shells with one to eight electrons. The Periodic Table lists all elements' atomic properties in Groups I through IIX, corresponding to between one and eight-valence shell electrons. In semiconductors "valence" and "conduction" refer to two distinct bands. In metals, though, these terms are used interchangeably. Group IV elements, such as silicon (Si), have four valence electrons. GaAs and GaP materials are Group III-V compounds from their column's position in the Chemical Periodic Table. Infrared light emitters diffuse an element, such as zinc, into a GaAs compound.

Elements with one electron in their valence shell are good conductors. Silver, gold, and copper are Group I elements, and good conductors of both heat and electrical current. A good conducting element with just one valence electron requires very little energy to remove this electron from the valence shell and set it free. Conversely, elements with eight valence electrons require much more energy to set a single electron free. Iron, cobalt, and platinum, Group VIII elements, are poor conductors and have eight valence electrons.

Metals, Insulators and Semiconductors

Solid-state physics divides solids into two major groups: *metals* and *insulators*. Metal conducts electric current under the presence of an electrical field. Under the same conditions, no current flows in an insulator. *Energy bands* are an untraditional criterion for judging metals and insulators. A completely filled band carries no electric current, even under the presence of an electric field.

Semiconductors fall between metals and insulators, having predominant characteristics within a small gap between the valence and the band immediately above. You can easily thermally excite electrons from the valence to the next higher band. Both bands contribute to the electric conduction, and both are only partially filled. Generally, a substance behaves as a semiconductor when its energy gap is 2.4 **eV** (electron volts) or less at room temperature. By contrast, carbon (diamonds) has a gap of 7 eV. The conductivity of a semiconductor is small compared to that of an insulator.

The complex forces associated with the repulsion or attraction of electrons are functions of neighboring atoms and the rapid motion occurring within an atom's shells. However, there are discrete energy levels associated with each shell. If the addition of energy in the form of heat, light, or an electric field disturbs the structure, the elements gain enough energy to move to a higher level within the atom. If you apply insufficient energy, the electrons remain within their original level and don't change states.

Assuming that there is sufficient energy for an electron to change states, the energy level that is losing the electron needs a replacement electron to complete its shell. It therefore develops an *attraction force*. It may capture its original departed electron, or it may capture another electron from the external energy source. In any case, when it does capture an electron, it releases *radiant energy*. The wavelength of this radiation depends upon the change in energy, which depends on the atom's structure. In some materials, this is broad spectrum electrical noise, while in other materials, it is visible light. The color of this light is wavelength-dependent.

Applying LED current combines electrons and holes, which effectively annihilates them as charge carriers. Photons are the released energy. Different semiconductor materials have different absorption and emission characteristics.

Atoms or molecules that are absorbing light have energy which excites their structure to "bump up" to a higher energy level. This excitation is wavelength-dependent. *Ultraviolet* (UV) light bumps electrons into a higher orbit. Vibrations are excited by infrared (IR) light. Rotations of electrons are excited by microwave energy while, conversely, emission is the decay back to a lower energy orbital and the ensuing radiation (emission or luminescence). Atoms excited by high temperatures are commonly called *atomic* or *optical* emissions. Exciting atoms with light is called *atomic fluorescence*.

You can tailor the desired light-emitting characteristics and color of LED devices made with a gallium arsenide (GaAs) or gallium phosphide (GaP) substrate, and an epitaxial layer (a single crystal, grown on the substrate).

The Nature of Light Emitters

Minority carriers recombine with majority carriers in a semiconductor, releasing kinetic energy, light, and heat to other carriers. The most effective way of introducing minority carriers is by carrier injection in a P-N junction. This is the basis of semiconductor light sources.

Any forward biased P-N junction emits some light; however, you must optimize certain key factors to maximize photometric efficiency. For example, light output is drastically reduced when reflection losses, internal absorption, or competing nonradiative recombination processes occur. Another type loss is *Fresnel's loss*. This is when light passes through two different media with different indices of refraction. You lose a portion of this light by it reflecting back at the medium interface. Selecting the right material in a state of high purity and crystal perfection, with proper doping, minimizes these losses. Proper geometrical structures also help reduce absorption and reflection losses.

The Absorption Process

The fundamental absorption process occurs when electrons move from the valence to the conduction band. In this process, an electron absorbs a photon and jumps from the valence to the conduction band. In the transition, you must conserve the total energy and momentum of the electron-photon system. The phonon or heat energy is small — about 0.05 eV or 1/20th that of a photon.

Material Selection

In an indirect-gap semiconductor, there is a difference in momentum between the initial and final states of a band-to-band transition. A photon, co-operating in the electron-hole recombination, must absorb the momentum. This requires a three-body collision; hence, the low probability of light emission from indirect gap materials. Conversely, vertical band-to-band radiative transitions in a direct gap material, such as GaAs, are more likely to produce light.

By growing single crystal-mixed compounds such as GaAsP and GaAlAs, it is possible to fit a material to a desired wavelength by varying the composition. As an example, you could have energy gaps of approximately 1.4 eV (electron volts) for GaAs to 2.26 eV for GaP. However, both GaP and GaAlAs are indirect gap materials, and in both alloy systems, light emission is possible only down to 650 nm. Other possible light-emitting materials, especially the II-VI groups (CdS, ZnTe, etc.), have band gaps large enough to produce blue light; however, they are very inefficient.

Types of Emitters

An emitter must be an efficient transmitter of the light generated internal to the light-emitting device. One problem is that the interface between two media causes some refraction. This refraction prevents much of the light from reaching you.

The following equation describes the fraction of light reaching the front surface:

$$F_c = (1 - \cos \Theta_C)$$

Solving the above equation yields 16.127°; therefore, 1 - cos 16° is equal to 3.9%. If you don't use an anti-reflective coating, you lose an additional amount of light.

There are GaAs emitters in which light generation takes place almost entirely in the P-region, and has only a slightly longer wavelength than the absorption edge of N-type material. You can bring out light through the N-region without excessive absorption, if it is not too thick. However, to eliminate internal reflection and dramatically increase

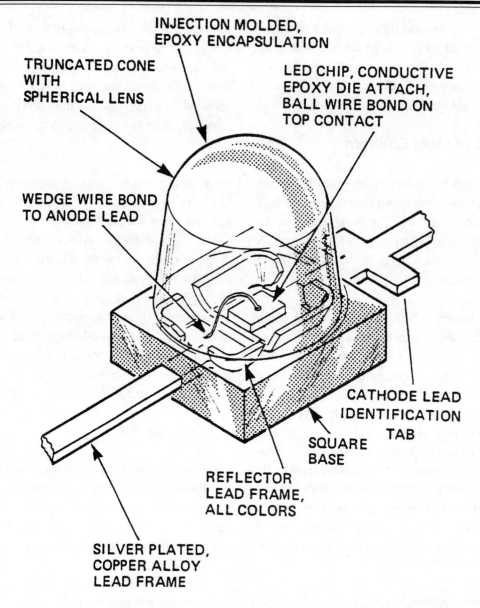

INJECTION MOLDED,
EPOXY ENCAPSULATION

TRUNCATED CONE
WITH
SPHERICAL LENS

LED CHIP, CONDUCTIVE
EPOXY DIE ATTACH,
BALL WIRE BOND ON
TOP CONTACT

WEDGE WIRE BOND
TO ANODE LEAD

CATHODE LEAD
IDENTIFICATION
TAB

SQUARE
BASE

REFLECTOR
LEAD FRAME,
ALL COLORS

SILVER PLATED,
COPPER ALLOY
LEAD FRAME

Figure 2-2. *Construction of an LED which uses a plastic dome lens.*
(Courtesy of Hewlett-Packard)

light output, you must place a semi-hemispherical plastic dome over the junction. (See *Figure 2-2.*) If the ratio of the dome-to-junction diameter is at least as great as that of the index of refraction of GaAs-to-air, none of the light reaching the surface will exceed the critical angle. (See *Figure 2-3.*) In actual practice, the thickness of the plastic in the domed lens reduces the light output by an improvement factor of 10 to 1 versus the theoretically attainable 25 to 1 ratio. You can make another improvement by using epoxy domes with an index of refraction of 1.7. This results in a threefold improve-

ment. Ideally, the index of refraction of the LED's coating should be:

$$n_1 \cdot n_2$$

$n_1 = 1$, the index of refraction for air, and $n_2 = 1.7$, the index of refraction for the plastic.

To achieve even greater efficiency, you can use Si (silicon) as the dopant in the P and N-regions. This shifts the wavelength several hundred Ångstroms, making it longer and reducing dome absorption.

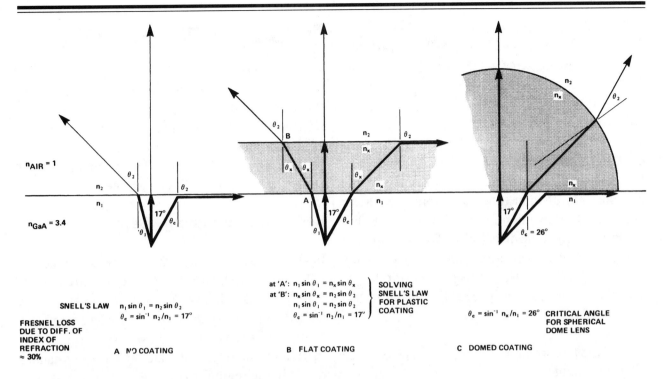

Figure 2-3. The critical angle.
(Courtesy of Hewlett-Packard)

Light Emitter Fabrication

It is common practice to coat the surface of an LED with SiO_2 (silicon dioxide), which has an index of refraction of 1.5, or with materials with higher indices of refraction, such as SiO (silicon oxide) or Si_3N_4 (silicon nitride). You can minimize reflectance loss using a coating of an odd number of quarter wavelengths in thickness. Discrete packaged single LEDs normally have two primary DC electrical parameters. These are *forward voltage drop* and the *breakdown voltage*. The forward voltage drop is of primary importance. GaAs requires approximately 1.2 volts. GaP requires in excess of 2 volts. Reverse breakdown voltages relate to the level of doping within the semiconductor, and range from 5 to 25 volts.

Infrared (IR) LED emitters exist as useful devices, and the following equation describes their external quantum efficiency, n_e:

$$n_e = \frac{(\text{Å})\lambda}{12,400} \bullet \frac{P(mW)}{I_f(mA)}$$

Typically, IR emitters have quantum efficiencies of 0.1 to 1% with premium IR emitters having somewhat higher ratings. P is total radiant flux, I_f is the LED's forward current, and l is the wavelength of the LED's light.

Luminescence

E. Wiedemann introduced the term *luminescence* in 1889 to describe light emissions that are not solely temperature induced. You can classify luminescent sources by the manner in which a source generates light. Three of the most common are *fluorescence*, *phosphorescence*, and *injection luminescence*.

Fluorescence
Applied energy in fluorescent sources raises an electron to a higher energy level, where it remains for approximately 10 nanoseconds. Light occurs after the electron returns to its original energy level.

Phosphorescence
This holds electrons in an intermediate energy level above their normal state. Light is emitted as elec-

29

trons gradually fall back to their original lower energy state. The most common example of this application is analog wrist watches with glow-in-the-dark dots next to the numerals.

Injection Luminescence

This the phenomenon in which we are most interested, where semiconductor material is concerned. It is the basis of LED operation. Lossew first noted the emission of light (photons) from a naturally occurring P-N junction in 1923. In 1962, studies of GaAs revealed the feasibility of achieving relatively high levels of electroluminescence emission from P-N junctions. The GaAs junction diode was appealing because it met three previously unfulfilled basic needs. These needs were:

1. Low level currents and voltages which produce useful light output.
2. A precise definition of the light-emitting area through semiconductor photolithography.
3. Switching these devices at very high speeds.

Naturally Occurring Luminescence

Here are two examples of naturally occurring luminescence. Every September, the beaches in California accumulate huge masses of bacteria which turn the surf into a dark, glowing red color. In New Zealand, millions of tiny glowworms glow from the roof of a deep, dark cave. Their generated light attracts insects upon which they feed.

Semiconductor Energy Gaps

In semiconductor materials, as is true with all crystalline solids, electrons can assume only certain energy levels. The two highest energy levels are the *valence* and *conduction* bands. The energy gap is the separation between these two energy bands. In a pure semiconductor material, electrons cannot reside in the region between the valence and conduction bands. Impurities allow electrons to reside in this otherwise forbidden zone reserved for absolutely pure semiconductor materials. Impurities

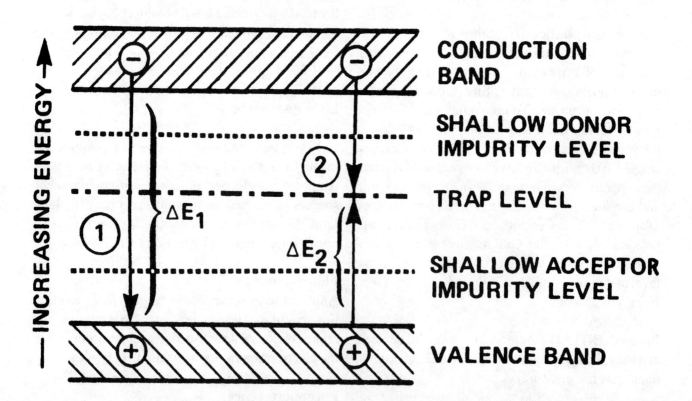

Figure 2-4. *The energy bands and relative donor and acceptor levels.*
(Courtesy of Hewlett-Packard)

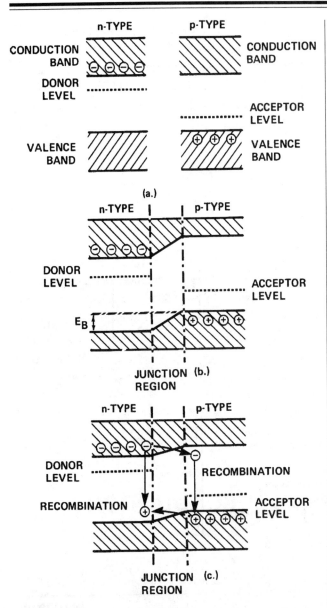

Figure 2-5. Recombination in the P-N junction. (Courtesy of Hewlett-Packard)

equilibrium. This is slightly less than the energy gap. *Figure 2-4* represents *before* and *after* reaching this equilibrium. Note the natural barrier which makes it difficult for an electron to cross the junction from the N to the P region. Applying an external bias overcomes this built-in potential, allowing injected holes and electrons to cross the junction. This recombination allows current to flow with an accompanying photon (light) emission.

Recombination

The injected electrons or carriers are annihilated by a carrier of the opposite type by one or two main recombination processes. These processes are either *radiative* or *nonradiative*. The radiative process releases photons and light. The nonradiative process releases phonons or heat alone. The radiative process occurs in two ways. First, band-to-band recombination is the direct recombination of an electron near the bottom of the conduction band, with a hole near the top of the valence band. (See *Figure 2-5*.) The photon energy produced is approximately equal to the bandgap energy of the crystal, and predominates in **direct bandgap** materials such as GaAs.

Indirect bandgap materials, such as GaP, experience radiative recombination by a **bound exciton**, forming and annihilating electron hole pairs at an "isoelectronic" center. A bound exciton occurs when an electron and a hole attract each another; however, they form a **bound** state in which the two particles revolve around each other. When annihilation occurs, it yields a photon with energy equal to the bandgap energy, minus an energy approximately equal to the binding energy of the exciton center. The following equation describes this:

$$\lambda = \frac{1240}{\Delta E}(nm)$$

ΔE is the energy transition in electron volts.

which add electrons to the conduction band (donors) yield N-type conductivity. Impurities which cause electron vacancies (holes) yield P-type conductivity.

An LED's P-N Junction

You can form a P-N junction by doping adjacent semiconductor materials with *donor* and *acceptor* atoms. Electrons and holes then flow across this junction without applying any external bias. A built-in potential barrier (E_B) occurs after reaching an

MATERIAL	BAND GAP ENERGY	EMISSION λ nm	TRANSITION TYPE
	eV		
Ge	0.66	1880	INDIRECT
Si	1.09	1140	INDIRECT
GaAs	1.43	910	DIRECT
GaP	2.24	560	INDIRECT
GaAs$_{60}$P$_{40}$	1.91	650	DIRECT
Al Sb	1.60	775	INDIRECT
In Sb	0.18	6900	DIRECT
Si C	2.2-3.0	563-413	INDIRECT

Table 2-1. *Bandgap energy emission in nanometers.*
(Courtesy of Hewlett-Packard)

Semiconductor Materials Used in LEDs

The previous equation describes the wavelength of photon emission related to the energy differential during a recombination process of an electron. To verify the equation, multiply the bandgap energy of any of the listed materials by its emission in nanometers. (Refer to *Table 2-1*.) You should have a constant product of 1,240 regardless of the material. Let's try GaAs$_{60}$P$_{40}$ — that would be 1.91 times 650, which equals 1241.5, or very nearly 1240.

The materials listed in *Table 2-1* range from infrared to ultraviolet with visible light in between. While these materials present theoretical opportunities at photon emission, practical constraints limit their uses. Some materials cannot have their P-N junctions doped, while others produce light outside a band of interest, or produce light so inefficiently that they are useless. Until recently, only GaAs, GaP, and the ternary compound Ga(As,P) were useful. Now we have a new material, AlGaAs.

Direct vs. Indirect Bandgap Material

Figure 2-6 illustrates the concept of indirect and direct bandgap transitions. This plot is of the material

GaAs$_{1-x}$P$_x$, with x = the mole fraction at various values. A *mole fraction* is the relative number of molecules in a given sample versus how many it could hold at its maximum capacity. There are two dips in the curves. One is designated as a **direct** minimum, and the other is designated as an **indirect**

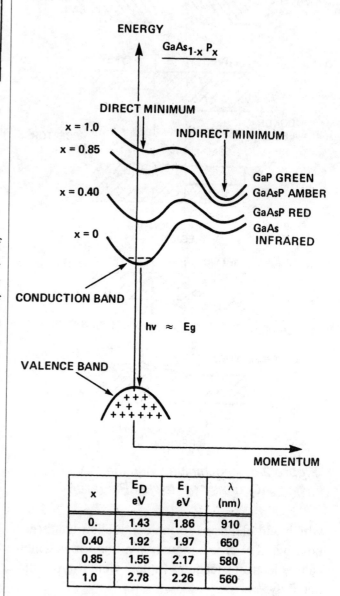

x	E$_D$ eV	E$_I$ eV	λ (nm)
0.	1.43	1.86	910
0.40	1.92	1.97	650
0.85	1.55	2.17	580
1.0	2.78	2.26	560

Figure 2-6. *Momentum versus bandgap energy for various GaAs and GaP compounds.*
(Courtesy of Hewlett-Packard)

minimum. Electrons in the conduction band will generally occupy states in the lowest energy minimum, and holes will occupy states near the valence band maximum. Both types of these electrons have equal momentum; however, electrons in the indirect minimum have different momentum.

Physics mandates that momentum must be conserved; therefore, band-to-band transitions may occur with high probability for electrons in the direct minimum. The probability of a band-to-band transition for an electron in an indirect minimum is almost zero. This is due to a third component (phonon or heat) which must also participate in this process to conserve momentum.

GaAs and $GaAs_{1-x}P_x$, up to x, approximately equal to 0.4, are primarily direct bandgap materials. This contrasts $GaAs_{1-x}P_x$ with "x" greater than approximately 0.4, as well as GaP. In these indirect gap materials, bandgap energy increases with an increasing x. (See *Figure 2-6* again.) Therefore, GaP is an inefficient light producer — it does not have special recombination centers, and the dominant recombination processes are non-radiative (just heat is produced). GaP has an effective wavelength of 560 nm. Refer to the CIE curve (*Figure 1-2* in Chapter 1), noting yellowish-green, the most easily-perceived visible color. The inefficiency of GaP as an indirect bandgap material is therefore partially compensated for by the uniqueness of the human eye as a photometric receiver of visible light. Nonetheless, you can enhance photon efficiency in these highly-visible, colored light-producing GaP materials.

Enhancing Photon Emission in Indirect Gap Materials

If we could enhance radiative recombination in indirect gap materials, such as GaP, it would enhance photon emission efficiency. Adding the appropriate impurities to form isocenters makes the trapped electron highly localized, diffusing the momentum

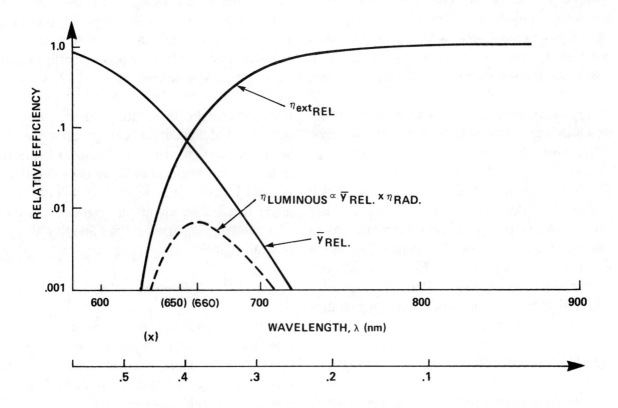

Figure 2-7. *Luminous efficiency of the human eye versus* $GaAs_{1-x}P_x$.
(Courtesy of Hewlett-Packard)

at its center, and increasing the probability of direct recombination.

In GaP, there are two types of isoelectronic centers. One results when a nitrogen atom replaces a phosphorous atom in the lattice. Another type forms by replacing an adjacent gallium and phosphorous pair of atoms with a zinc-oxygen pair, which must have the same number of valence electrons. For N-doped GaP, bound exciton recombination stimulates emission at 565 nm (green). For Zn-O doping, the corresponding wavelength is 700 nm (red). Nitrogen doping in indirect-gap $GaAs_{1-x}P_x$ produces relatively high quantum efficiency red and yellow light.

Quantum Efficiencies of LEDs

Efficiency is a measure of the ratio of electric energy required for a given output of photometric (visible light) energy. Ideally, you should seek an electroluminescent device with a high efficiency; however, in practice, this is only one of several factors determining the usefulness of the device. Photons may be streaming out quite well, but they must be coupled to a detector. Two factors influence this. One is the reabsorption of the photons back into the emitting material. There is internal reflection at the crystal/air interface. This also reflects the photons back into the crystal and subsequently the crystal reabsorbs them.

The next concern is matching the output wavelength to that of the detector. Assuming it is a human's eye, people perceive light in the 400 nm to 700 nm range. However, a silicon photodetector's range is wider (400 nm to 1,100 nm). *Figure 2-7* is a plot of the human eye's efficiency, yREL, versus the n_{ext} REL efficiency. Note that the point at which they cross is 655 nm, or red.

Semiconductor Materials Processing

The processes used in manufacturing LEDs are all refinements of silicon processes used to make transistors and other semiconductors. These processes consist of *crystal growing*, *epitaxial deposition*, *controlled impurity diffusions*, *photolithography*, and *vapor deposition* of very thin films of materials.

The LED's Structure

It is nearly impossible to achieve a single crystal with the proper GaP concentration, since to Ga(As,P) is not a congruently (harmoniously compatible) melting material. Therefore, manufacturers produce Ga(As or P) as an epitaxial layer grown on a substrate of either GaAs or GaP. The substrates are wafers sliced from ingots grown through the Czochralski process. These mechanically and chemically polished wafers produce a crystal substrate virtually free of lattice imperfections.

Transparent vs. Opaque Substrates

The photons emitted at the junction of a P-N electroluminescent diode scatter randomly. If the diode's substrate is opaque (such as with GaAs), only those photons emitted upward through the light pipe (the path from the substrate to the outside world including the air-plastic interface), within the critical angle defined by Snell's Law, will be useful light. (See *Figure 2-8*.) Snell's Law states that a ray of light, propagating within a medium of refractive index n_1, approaching the boundary of a second medium with a refractive index n_2, must be at the critical angle in order to pass through the boundary. If it is at an angle less than the critical angle, it reflects back into its originating medium. The following equation defines Snell's Law as the critical angle Θ:

$$\Theta = \cos^{-1}(n_2/n_1)$$

GaP is nearly transparent compared to GaAs. Diodes that form in an epitaxial layer grown on a GaP substrate exhibit improved efficiencies, because GaP absorbs less photons than GaAs. There are now external light pipes.

LIGHT RAYS FROM LED TRAVEL THROUGH
LIGHT PIPE TO DIFFUSING LAYER. THE
DIFFUSING LAYER SPREADS LIGHT TO
FROM AN EVENLY LIGHTED RECTANGULAR
SURFACE

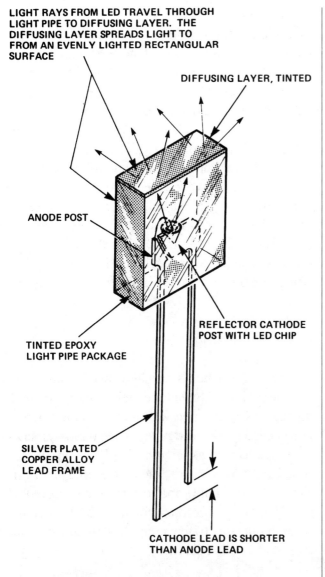

DIFFUSING LAYER, TINTED

ANODE POST

REFLECTOR CATHODE
POST WITH LED CHIP

TINTED EPOXY
LIGHT PIPE PACKAGE

SILVER PLATED
COPPER ALLOY
LEAD FRAME

CATHODE LEAD IS SHORTER
THAN ANODE LEAD

Figure 2-8. *An LED showing its
plastic light pipe.
(Courtesy of Hewlett-Packard)*

Temperature Effects on LEDs

The effective energy gap in both indirect and direct gap semiconductors tends to decrease with temperature. This results in small increases in the emitted wavelength. For direct gap semiconductors, it is approximately 0.2 nm/°C. The nitrogen-doped indirect bandgap emitters are less temperature dependent, with a 0.09 nm/°C positive variation in wavelength.

The radiant energy of an LED decreases with temperature, approximating -1%/°C. The responsiveness of the human eye to this semiconductor phenomenon, in the red region of 650 nm, changes approximately -4.3%/nm. In the green region of 565 nm, it changes at a diminished rate of -.86%/nm.

Band Structure

A semiconductor is a substance in which the highest occupied energy band is completely full at absolute zero, T = 0 K or -273.15°C. However, at room temperature, because the gap above this band is small, electrons may be thermally excited above the valence band into the next higher band, the **conduction** band. When excited, electrons cross this gap, causing electrons to occur at the bottom of the conduction band and holes to occur at the top of the valence band. Only the valence and conduction bands, though, are of interest to us since bands lower than the valence band are completely full, and bands above the conduction band are completely empty; therefore, neither band contributes to current flow.

Semiconductor Processing

Controlled diffusion is when you place a chip of the substrate material in a chamber under a steady concentration of the desired impurities. You maintain these impurities in a gaseous state surrounding the chip. The impurities diffuse into the chip as you raise the temperature. The depth of the penetration depends on the temperature, the duration of the process, and the impurities you use.

If you require a new and different layer, like transistors do, the new layer forms from being diffused on top of the old layer, but to a lesser depth. However, the new layer still has the old impurities embedded in it. Repeating this process several times yields layers with concentrations of different impurities, and increases conductivity. Semiconductor manufacturers developed the epitaxial growth method to avoid this.

Epitaxial growth involves depositing layers of desired impurities onto the chip by placing it within a gaseous reaction system chamber. The Si or Ge layers precipitate so slowly and gradually that you can maintain the crystalline continuity between the chip and the new layers. *Ion implantation* is another technique. This shoots impurities toward the surface of the semiconductor after a static accelerator accelerates them. The depth of penetration depends on the acceleration potential with a few kV, a typical value, and 10 nm (a typical penetration depth).

Crystal Growth

The majority of LED products have a GaAs substrate. The single crystals necessary for reproducible device manufacturing are prepared by either a *Bridgman* or *Czochralski* method.

The Bridgman Method

This method uses a seed of single-crystal GaAs, previously prepared and oriented in the appropriate crystalline direction. A "boat" holds the seed, which has a semi-cylindrical cross-section. One end of a quartz reaction vessel holds the boat, which you place into highly pure gallium and arsenic at 1250°C. The gallium and arsenic combine, and with the continued application of this heat, become a homogenous liquid. As you gradually remove the heat from the seed end of the vessel, a crystal begins to grow onto the seed. If all parameters are correct, the new crystal assumes the orientation of the seed. To give the crystal the desired electrical properties, you dope the material with special impurity atoms; however, unwanted impurities are less than one part per million.

The Czochralski Method

In the Czochralski process, you separately synthesize the polycrystalline GaAs material using the Bridgman technique. You then place the material in a quartz cup covered with liquid glass. An overpressure exists over the inert gas, preventing it from breaking down into separate elements. Lower a seed crystal (prepared in a manner similar to the seed in the Bridgman process) on a rotating rod to the surface of the molten GaAs. After it reaches the proper temperature, slowly extract the seed. This causes a crystal to grow onto it. Controlling the temperature of the liquid GaAs and the speed with which you raise the seed determines the new crystal's diameter and shape. The desired electrical properties again result from doping. GaP crystals form by the Czochralski process in much the same way as GaAs crystals; however, the liquid temperature and the overpressure must be greater. The benefit of the Czochralski process is that it grows larger and longer crystals. The drawback is that it requires a more skilled and experienced operator. The Bridgman process is more appropriate for automation.

LED Processing

After the crystal cools, visual inspection and measurement of electrical properties occur. A diamond precision saw, aligned with a laser to the orientation needed for the next process step (epitaxy), saws the ingot, as well as removing a slice from the exposed end. An x-ray diffractometer examines this slice to verify that proper orientation occurred. If the orientation is correct, you then thinly slice the entire ingot (substrate wafers) for thickness, parallelism, and polish quality.

Epitaxy

One of the most critical steps in the manufacturing process is depositing an epitaxial layer onto each polished GaAs or GaP substrate wafer. This layer is an alloy of GaAs and GaP (or new alloys), and its composition determines the color.

To grow the epitaxy layer, place the wafer in a chemical vapor deposition reactor, through which gases flow. Next, heat the wafer in hydrogen. Hydrogen chloride gas etches it to clean its surface to an extent far greater than would be possible by mechanical devices in air. You then deposit very pure atoms of arsenic, phosphorous, gallium, and dopant material on the wafer in a manner duplicating the single-crystal nature of the wafer. Varying the composition of the alloy attains the desired final composition. The dopant deposits the correct amount of impurity atoms in the layer. After depositing the layer, test the epitaxial wafer to ensure it emits the correct color of light. If acceptable, the wafer is the starting point in the fabrication process.

Wafer Fabrication

The first step in wafer fabrication is growing a layer of silicon nitride over the wafer's surface. Outline windows on the wafer, using photolithographic techniques, and diffuse the necessary impurities into the surface at high temperatures. The shape of the junctions formed depends on the type of device being manufactured. The nitride layer not only defines the junctions, or dice, but also provides passivation to ensure long-term reliability and high photometric efficiency.

To make contact with the dice, you need to photolithographically deposit aluminum patterns on the wafer. This laps the wafer from its back side to a final thickness, depending on the required die size. You then evaporate gold and germanium onto the back of the wafer, and alloy it into the substrate in a furnace. Applying a final layer of gold provides the best surface for subsequent steps.

Scribed lines between the dice on the wafer occur after further visual inspection and electrical testing. These lines, along the crystallographic planes, serve as fracture lines. Applying a small strain breaks the wafer into individual dice along these scribed lines, just like breaking window pane glass. Bonding the dice onto headers, lead frames, or PC boards forms a strong mechanical connection and provides good electrical contact. Bonding wires to the frontside metallization creates electrical contact between the device and the package leads. Packaging the encapsulation follows, including attaching a light pipe and lens to the lead frame or PC board.

Encapsulation Materials

Plastic die encapsulation materials must exhibit four primary characteristics:

1. A high degree of light transmission.
2. Good adherence of the material to the die, and moisture resistance.
3. Thermal expansion coefficients approximating those of other materials in the device. Various materials expanding at differing rates will cause strains and possible eventual separation at the junctions of dissimilar materials.
4. Handling and curing properties that lend themselves well to mass production techniques.

Meeting all of these characteristics is difficult, due to a limited choice of materials. Also, digits using internal reflecting cavities or light pipes require plastic parts that either have intrinsically high reflectance, or can be plated to obtain high reflectance. This often eliminates fillers. *Fillers* are epoxies and silicones added to the encapsulations of other electronic components to tailor expansion rates, viscosity, and adhesion properties. This promotes compatibility of the various pieces and improves maximum temperature ratings. LED lamps are limited to $100\,^{\circ}C$, and LED digits (e.g. seven-segment displays) to $85\,^{\circ}C$.

AlGaAs

Hewlett-Packard, the world's largest electronic test equipment and printer manufacturer, has a new red light-emitting diode (LED), based on the aluminum gallium-arsenide (AlGaAs) material, with dramatically improved efficiency and less light degradation over time. It is only a little more expensive than ordinary LEDs.

All previous LEDs were homostructures, consisting of P-N junctions formed in one type of material. All optoelectronic devices, such as LEDs, prepared their materials the same way. The major differences were assembly processes and final device fabrication materials. LEDs now have several layers of materials with different bandgaps (heterostructures). These LEDs are more efficient due to single-sided injection (more "focused" or directed light emission) and reduced internal absorption.

In homostructures, under forward bias, electrons inject themselves into the P-type material, and holes inject themselves into the N-type material. Some fraction of these carriers recombines with the majority carriers on the P and N sides of the junction, and emit the near-bandgap light characteristics of the LED. The radiative efficiency on the P and N sides is usually quite different. Generating more light is possible by reducing or eliminating minority-carrier injection into this less radiatively efficient material.

We can eliminate this undesired injection if the P-N junction occurs at the interface between two materials of different bandgaps. You may change the bandgap by altering the alloy composition in the AlGaAs system (changing the ratio of aluminum to gallium in the compound). *Figure 2-9b* is an energy diagram for this, and *Figure 2-9a* shows the homojunction structure for comparison. The discontinuity in the valence band adds much more to the hole potential barrier, than the conduction band discontinuity adds to the electron potential barrier. *Figure 2-9b* shows the band configuration for a single heterostructure (SH) device. It effectively eliminates hole injection into the wide-gap N-type material under forward bias. This provides single-sided injection.

The second advantage is less reabsorption inside the material than with homostructures. A further reduction occurs in reabsorption by placing a second wide-gap layer on the other side of the narrow-gap layer. (See *Figure 2-9c*.) This is a double heterostructure (DH) device. The second heterointerface prevents the injected electrons from diffusing out of the narrow-gap active layer. It also enhances the generated light's chances of escape; however, the narrow-gap and wide-gap materials must have very nearly the same lattice structure to avoid dislocations at the material interfaces. This would severely reduce the light-generation efficiency. Fortunately, in all AlGaAs systems, the lattice constant changes very little with alloy composition. This technique predominated in the last 15 years to fabricate high-efficiency infrared LEDs, and low threshold lasers, though not visible light LEDs — until now.

You can use the AlGaAs system to efficiently fabricate diodes that emit visible red light; however, this is a recent breakthrough due to the difficulty in growing the high aluminum content layers necessary. Liquid phase epitaxy (LPE) is the only technology providing quality material comparable to that of infrared LEDs. The breakthrough is due to LPE reactors now being able to grow large volumes of high-quality multilayer devices.

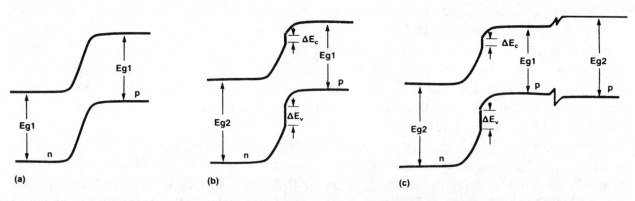

Figure 2-9. *Typical energy diagrams for (a) a P-N homojunction, (b) a single heterostructure (SH), and (c) a double heterostructure (DH) LED. (Courtesy of Hewlett-Packard)*

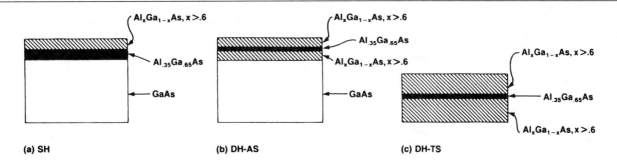

Figure 2-10. *The three types of commercially available AlGaAs LEDs.*
(Courtesy of Hewlett-Packard)

Blue LEDs

There has been a lot of research and interest in blue LED manufacturing. $GaAs_{1-x}$ and GaP, the normal semiconductor materials, are very impractical because the bandgap is too small. This limits the wavelength of the emitted radiation toward the lower end. In the 1970s, GaN (gallium nitride) was the semiconductor material of choice for blue LED research, but efforts proved fruitless and were abandoned. This was primarily due to the material's inability to be P-doped with sufficiently low resistance. The light in this semiconductor was not produced by the radiative recombination of injected charge carriers at the P-N junction, as is the case with the other type III - V materials. Light was inefficiently produced by accelerated electrons, generated in very high resistance "i" layers in a metal GaN-n-GaN layer by the collision-ionization process. Blue light also results from using ZnS and ZnSe, but it suffers from no-to-low resistance P-N junctions.

The Three Types of AlGaAs LEDs

There are three types of AlGaAs LEDs available. *Figure 2-10* illustrates the composition diagrams for all three. The first SH LED consists of at least two layers of AlGaAs on a GaAs substrate. The first layer is the active layer, made of $Al_{.35}Ga_{.65}As$. The top layer is a "window" made of AlGaAs, with the mole fraction greater than approximately 0.6. The light produced escapes out the top and sides of the epitaxial material; however, the substrate absorbs the light hitting it. Note that there are

chemical compounds with subscripts, such as .4 and .6. These will always add up to 1. Also, there are instances in which the mole fraction can be a number of different combinations corresponding to a graph. In these cases, the compounds have subscripts of x and 1-x, which will again add up to 1.

This type of device is the easiest to grow since it has the fewest layers, and thickness control is not critical. It is also the dimmest, and only marginally better than the best nitrogen-doped GaAsP (GaAsP:N) devices. Since it is the easiest to manufacture, it has been commercially available the longest.

Refer back to *Figure 2-9b*, which shows a DH chip. There are at least three AlGaAs layers on a GaAs substrate. The first layer is a wide-gap injecting layer, and the second is a 2-micron thick active layer. There is a thick wide-gap confining layer above the active layer. Unfortunately, the substrate again absorbs most of the generated light. We therefore call it the DH-AS, for *double-heterostructure absorbing layer*. It is still twice as bright as the SH AlGaAs LED. Manufacturing it requires much better thickness control and at least one layer of epitaxial material.

Figure 2-11 shows the relative brightness curves of the SH, DH-AS and the GaAsP:N LEDs. Note that at lower forward current levels, especially below 2 mA, the DH-AS is more than five times brighter than other LEDs. This allows many new low-current applications.

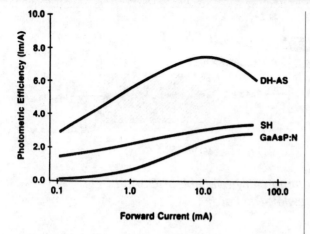

Figure 2-11. *Photometric efficiency in Lumens/ Amp of various LED technologies. (Courtesy of Hewlett-Packard)*

The last structure is the DH transparent or DH-TS AlGaAs device. (See *Figure 2-10c*.) It is similar to the DH-AS chip; however, you grow one or both of the wide-gap layers several thousandths of an inch thick, and etch away the entire GaAs substrate. This eliminates absorption in the substrate and produces the greatest efficiency yet — more than twice as great as that of the DH-AS LEDs. This technology is only used where applications can justify the cost. It is difficult and expensive to manufacture due to the growth of the thick AlGaAs layer, which is technically very difficult to achieve.

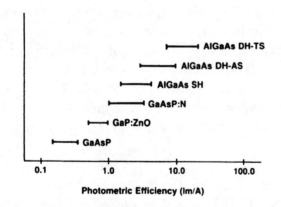

Figure 2-12. *Photometric efficencies of various plastic domed LEDs. (Courtesy of Hewlett-Packard)*

Comparisons of Other LEDs

The GaAsP alloy system is the basis for most commercially-available LEDs. You may subdivide these into two main groups: the *nitrogen-doped* and the *non-nitrogen-doped* LEDs. *Figure 2-12* shows the structure of the non-nitrogen doped LEDs (here simply referred to as GaAsP). This represents the dimmest and first LED technology. The radiometric efficiencies of these various LED technologies, at 20 mA forward current, cover almost a 100:1 ratio.

Growing GaAsP epitaxial material with a slightly larger bandgap on a GaP substrate (which is transparent to the light generated), and doping both sides of the P-N junction with nitrogen, allows you to realize an order of magnitude greater photometric efficiency. LEDs made this way are labeled GaAsP:N. You grow both this and the earlier dimmer version LEDs by *vapor phase epitaxy* (VPE), which lends itself well to mass production.

Somewhere between these mediocre and superperforming LEDs is the ZnO-doped GaP (GaP:ZnO), which grows in a two-step *liquid phase epitaxy* (LPE) process. An N-type, ZnO-doped

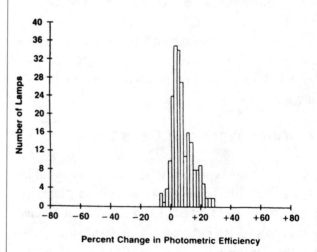

Figure 2-13. *A histogram of the photometric effiency of 215 DH-AS LEDs after 1,000 hours of operation. (Courtesy of Hewlett-Packard)*

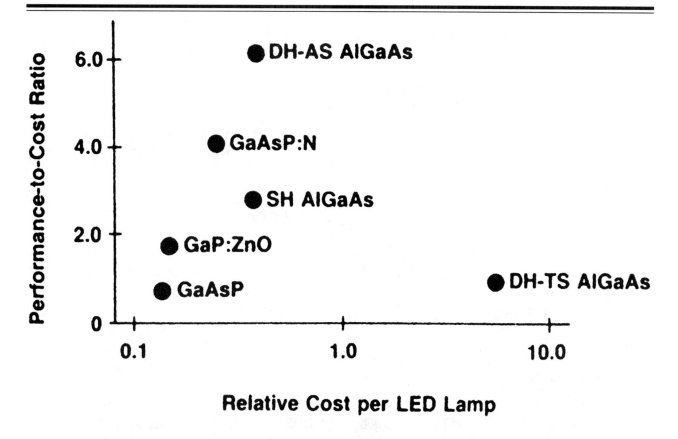

Figure 2-14. *Typical performance-to-cost ratios of various types of red LEDs.*
(Courtesy of Hewlett-Packard)

layer grows on a GaP substrate. Then, a P-type Zn-O layer grows on top. All of the red light emission from this material comes from the recombination at the ZnO centers on the P side of the junction. This material is radiometrically efficient (>10% external quantum efficiency for an encapsulated chip). The peak response, though, is approximately 700 nm — this is where the human eye response is fairly weak. Another drawback that is ZnO centers quickly saturate as the forward current through the junction exceeds 2 mA. This dramatically reduces radiometric efficiency.

SH and DH AlGaAs lamps operate at significantly lower voltages than GaAsP:N and GaP:ZnO. This is due to lower substrate and contact resistances, and a slightly smaller bandgap. This LED technology is best used for multi-lamp configurations and laptop PCs, where power consumption is critical.

The most interesting phenomenon, though, is a subtle one you would not immediately notice. Degradation characteristics of GaAsP and GaAsP:N are good. After 1,000 hours of operation at maximum drive, they degrade by approximately 15%; however, AlGaAs LEDs actually increase in radiometric efficiency as they age. *Figure 2-13* is a histogram showing percentage change in efficiency for 215 Hewlett-Packard lamps subjected to 1,000 hours of 30-mA 55° C stress. The primary reason light output degrades over time is that impure atoms, such as interstitial copper or zinc, migrate to positions in the lattice where they act as nonradiative recombination centers.

Figure 2-14 summarizes LED performance-to-cost ratio. DH-TS AlGaAs is the highest performance technology; however, manufacturing costs restrict applications. When "power signaling" products mature (automotive tail lights, airport marker bea-

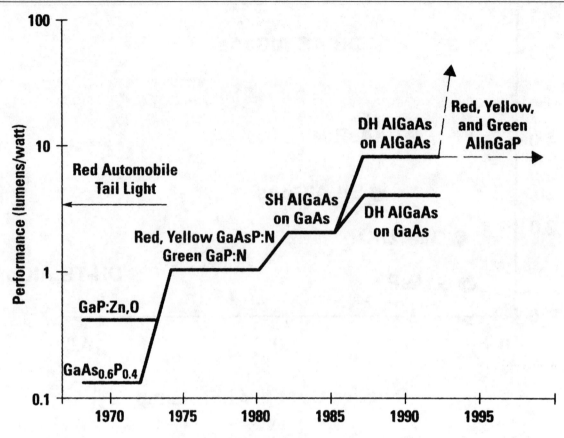

Figure 2-15. *The evolution of LED technologies.*
(Courtesy of Hewlett-Packard)

cons, traffic stop lights, etc.), intense red light DH-AS AlGaAs technology will serve it well. Previously, using LEDs in these applications was hopeless. *Figure 2-15* shows the evolution of LED technologies, with the attainment of a bright-enough LED for use in automobile tail lights only occurring at about 1990. Now, the real challenge comes in trying to reduce manufacturing costs so it is an economically viable candidate for these types of applications. *Table 2-2* explains how to properly rate an HP DH-AS LED at various viewing angles against other LEDs.

LED Ratings

Most LEDs are rated in terms of their on-axis intensity at a given drive current. The unit of measure for this parameter is the millicandella (mcd). This number, however, can be misleading when comparing different lamps because it is package dependent. There is a trade-off between on-axis intensity and radiation pattern; lamps made using chips that have the same efficiency can have dramatically different axial intensity (mcd) ratings depending upon the viewing angle. For example, a plastic lamp made using a typical HP DH-AS chip can have any of the mcd/viewing angle combinations listed in the table. For a given application, care must be taken to choose a lamp that has both the on-axis intensity and the viewing angle required.

On-Axis Intensity vs. Viewing Angle
HP DH-AS LED

On-Axis Intensity (mcd)	Viewing Angle (degrees)
1000	8
500	17
250	24
200	30
100	45
60	60

Table 2-2. *Rating a DH-AS LED at different viewing angles.*
(Courtesy of Hewlett-Packard)

Chapter 2 Quiz

1. Semiconductor phenomena strongly influence LED fabrication, T or F.
2. Optoelectronic emitter development preceded optoelectronic sensors, T or F.
3. An electron's outer or valance shell may have up to 8 electrons, T or F.
4. The Periodic Table describes chemical properties of the elements, T or F.
5. Solid state physics divides solids into two groups, insulators and metals, T or F.
6. Who first observed light striking a transistor's crystal structure and the ensuing photoconductive charge?
   A. Bohr.
   B. William Schottky.
   C. E. Wiedemann.
   D. Lossew.
7. If a substance has a 2.4 volt or less energy gap, it behaves as:
   A. An insulator.
   B. A metal.
   C. A semiconductor.
   D. A conductor.
8. The basis of semiconductor light is the introduction of minority carriers by carrier injection in a PN junction, T or F.
9. The fundamental absorption process occurs when electrons move from the conduction band to the valence band, T or F.
10. A light emitter loses efficiency because of:
    A. The interface between the two media, which causes refraction.
    B. Cloudy lenses.
    C. Diffused projection patterns.
    D. Applying too much forward current.
11. You can minimize reflectance loses by using a lens coating of an odd number of quarter wavelengths, T or F.
12. E. Wiedemann first used the term luminescence to describe light emission of a temperature induced nature, T or F.
13. The two most important parameters for a single visible LED are forward volt drop and reverse breakdown voltage, T or F.
14. You can minimize internal LED reflections by using:
    A. Tiny internal mirrors.
    B. Polishing the semiconductor better.
    C. Using a hemispherical plastic dome over the junction.
    D. All of the above.
15. The highest energy levels are the valence and conduction bands, T or F.
16. What is responsible for allowing electrons to reside in the otherwise forbidden zone between the valence and conduction bands?
    A. Heating.
    B. Introduction of a magnetic field.
    C. Adding impurities to the LED material.
    D. None of the above.

17. A bound exciton forms and annihilates electron hole pairs at _____ centers.
    A. Valence band.
    B. Conduction band.
    C. Isoelectronic.
    D. A and B.
18. An LED's efficiency is a measure of the ratio of electrical energy required to produce a certain output of _____ ?
    A. Light current.
    B. Resistance.
    C. Photometric (visible light) energy.
    D. Dark current.
19. The _____ process uses a seed of single-crystal GaAs oriented in the crystalline direction.
    A. The Bridgman process.
    B. The Czochralski process.
20. The _____ process requires a more skilled operator and is less well suited to automation.
    A. Bridgman.
    B. Czochralski.

Chapter 3
Measuring/Testing Optoelectronic Devices

Chapter 3
Measuring/Testing Optoelectronic Devices

This chapter examines measuring and testing a wide variety of both visible (photometric) and non-visible light (radiometric) optoelectronic devices. Be forewarned that optical measurements are not nearly as accurate as electrical measurements. The electro-optical test instruments with which we will perform the following nine tests are from Tektronix and International Light:

1. Testing visible LEDs.
2. Testing radiometric IR LEDs.
3. Correcting for the Law of Cosines (also known as Lambert's Law).
4. Using the proper spectral filters and detectors when measuring various optoelectronic devices.
5. Testing optical communications.
6. Testing photographic exposure settings.
7. Measuring pulsed light with an oscilloscope.
8. Measuring small areas of visible light.
9. Testing industrial CRTs in a TV station.

An Overview of Light Measuring Instrument Types

There are five basic types of instruments with which we make light measurements. These are:

1. Light intensity meter.
2. Photometer.
3. Radiometer.
4. Colorimeter.
5. Spectroradiometer.

The light intensity meter is by far the simplest and least expensive, and is used in photography and lighting measurements. Its spectral (color) sensitivity only roughly approximates that of the human eye or the response curve of the color film you are using.

What a Light Intensity Meter Does and Can Possibly Do

Be forewarned: the average light intensity meter receives unfiltered light that naturally does not comply with the color spectral response of the human eye, which the CIE curve represents. (See *Figure 1-2* in Chapter 1) This human eye-to-instrument discrepancy is precisely why you need a color temperature meter. The overall sensitivity of a light meter's light sensor, such as a silicon blue cell, is calibrated from either a 2856 K or 2870 K tungsten light source.

A light meter's basis of operation uses the additive photographic exposure system of exposure values. Note that all five quantities in the following basic equation have "v" subscripts for one of five different values, thus its name. This equation consists of:

$$E_v = [A_v + T_v] = [S_v + B_v]$$

The next equation states this *qualitatively* as:

Exposure Value (E_v)
= [Aperture Value (A_v) + Time Value (T_v)]
or
= [Speed Value (S_v) + Brightness Value (B_v)]

This equation defines the *aperture value* as:

$$(A_v) = \ln N^2 / \ln 2$$

N is the f/no. or *f* stop setting of your lens.

This equation defines the *time value* as:

$$(T_v) = \ln t^{-1}/\ln 2$$

t is exposure time in seconds, or fractions thereof.

This equation defines *speed value* as:

$$(S_v) = \ln 0.3S/\ln 2$$

S is the ASA (ISO) speed of your film.

This equation defines the *brightness value* as:

$$(B_v) = \ln Bfl/\ln 2$$

Bfl is the illuminated object's brightness in foot-Lamberts, and relates to B_v by this equation as:

$$Bfl = 2^{Bv}$$

This math, which may otherwise appear tedious to you, does have practical applications. The second equation of the next set of equations merely restates the basic *exposure value* to make this principle of deriving brightness level (B_v) from the exposure value (E_v) more obvious. The first equation restates *exposure value* as:

$$E_v = [(A_v) + (T_v)]$$
$$E_v = [(S_v) + (B_v)]$$

Our immediate goal is to use the second equation and find a way to make S_v equal zero. This would naturally greatly simply the second equation by allowing S_v to drop out of the equation's right member to just equal:

$$E_v = B_v$$

If we could somehow make E_v and B_v equal, it would allow you to legitimately directly read the brightness level (B_v) from your light meter's normal output — its exposure level (E_v). You **can** do this,

though, by setting the S_v or speed value to 3.1. Do not confuse the speed value (S_v) with the ASA (ISO) film speed. It is **not** equal to this under any condition except one, at an ASA (ISO) speed setting of 3.1.

You might feel uneasy about how we mysteriously (and not thoroughly) proved how we made $S_v = 0$. This all hinges on taking the natural log (ln) of 1; which, in this case, conveniently equals 0. Again, refer to the equation for S_v, and note it equals the natural log (ln) of 0.3S. By making the ASA (ISO) setting 3.1, this becomes ln 0.3(S) = ln 0.3 (3.1) = ln 0.93 = 0.0726. This is a very small number! Admittedly, if we fudge a little and make ln 0.93 into ln 1, it equals zero, and the fraction of (ln 0.3S/ln 2) has a zero in its numerator. Therefore, S equals 0. This eliminates it from the right member, which is how we realized our immediate goal. However, what do you do if your light meter does not go as low as ASA (ISO) 3.1? You may use an ASA (ISO) 100 film setting, and subtract five from the E_v reading to obtain B_v. You can use this same derivation method as an exercise to prove this for yourself.

Another handy calculation obtains luminance (photometric brightness, B) from illuminance (E) as: B = (k • E) (1/p). B is in candles/ft², and *k* is the reflectivity or percentage of received, non-reflected light. This percentage is always 1.0 viewed straight ahead, or less, if the light source is not directly in line with the light sensor. It equals the Cosine of the light's less-than straight-on angle of arrival. This is Lambert's Law.

The photometer measures visible light with far greater precision and accuracy than the light intensity meter. It very closely matches the spectral sensitivity of our eyes, and its peak sensitivity is in the yellow-green area, just like our eyes. You can make measurements in units of *illuminance, luminance intensity*, or *luminous flux*, depending upon the physical configuration of the detector (light sensor) you use.

The radiometer measures both photometric and radiometric light sources spanning beyond our visible spectrum on both ends. This instrument's spectral sensitivity is equal for all wavelength light, unlike our eyes and the photometer which mimics our eye's response to colors. You can make measurements in terms of *irradiance, radiance, radiant intensity,* or *radiant flux,* which are all analogous to the photometer's terms; however, its spectral sensitivity is flat, as we've said.

A radiometer would measure the true radiant energies that visible LEDs release, and determine their radiant light producing efficiencies. A photometer would only measure their visible light output; however, this is really a moot point, since we can only see visible light. This is an example demonstrating the differences in these instruments.

A photometer's optics control its sensitivity response, attempting to very nearly match that of the human **eye**. Another example of trying to overcome the logarithmic nature of another human sensor, the ear, is the *dBA convention* (not to be confused with normal dB output). This is a sound pressure level specifically biased or weighted for the human **ear**. This curve peaks at about 3.6 kHz.

A *colorimeter* measures the color of a light source or reflecting object. The numerically indicated color is in one of several color specification systems using two or three color coordinate values.

A spectroradiometer measures the quantity of light at each wavelength through your specified range. You can derive both photometric and radiometric terms by mathematically integrating the spectral

Figure 3-1. *Optical measuring instruments (photometer and a detector) from Tektronix. (Courtesy of Tektronix)*

data with an appropriate weighting factor, to correct for the eye's nonlinearity sensitivity which peaks in the green-yellow region of 565 *nm*. This instrument is also called a *scanning colorimeter* and is quite expensive, from $20,000 to $100,000. Most of these instruments have a dedicated computer with which they crunch the vast numbers to arrive at a reading. They also take readings on a wavelength-by-wavelength basis which eases the conversion of data from one set of unit symbols to the next.

Figure 3-1 shows an array of optical test instruments from Tektronix. Starting at the far left, the models J6503 and J6523 are luminance probes with 8° and 1° fields of view, respectively. A 1° field of view represents a spot diameter of 2.1 inches at 10 feet. The fields of view for both instruments are proportionately larger for greater distances and smaller for closer distances. The J6523 contains an optical viewing system so you can view the spot you measure through the eyepiece. These are admittedly older instruments; however, so many of them exist today in labs that they were selected as our representative samples.

The next two instruments are the J16 and the J6511. The J16 is a portable digital photometer/radiometer capable of making numerous light measurements. The J16 consists of the mainframe and one of eight detachable probes, which you can mount on either the J16 or an extension cable. These interchangeable probes measure illuminance, irradiance, LED outputs, and relative intensity. You do not need to recalibrate the instrument when you change probe types. All these probes use silicon photodiodes individually corrected with multi-element glass filters for maximum stability and accuracy.

The J6511 illuminance probe to the far right in *Figure 3-1* measures incident light in footcandles. You can multiply by the reflectance factor (to be explained) to obtain the luminance in foot-Lamberts. This probe is a cosine law-corrected instrument. This means that since the intensity of light diminishes at the cosine of the incident angle (or angle of arrival), irradiance and illuminance detectors

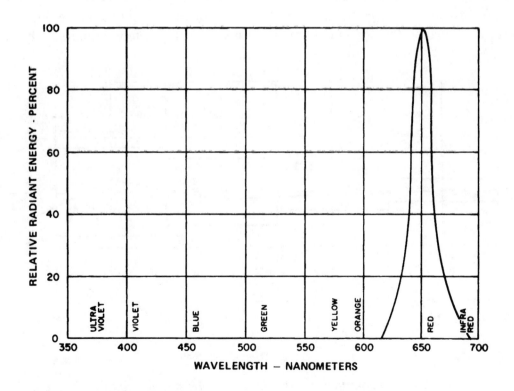

Figure 3-2. *A red LED's light response.*

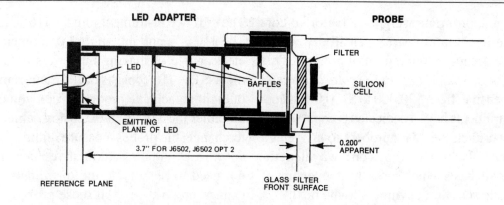

Figure 3-3. *An LED adapter to limit your measurement to just useful straight ahead light. (Courtesy of Tektronix)*

(probes) require correction. This is especially true when using detectors with optical filters.

Testing Visible LEDs

There are two immediate problems encountered when measuring a visible LED's output. First, the LED usually peaks in the red region of the light spectrum. Unfortunately, a standard photo-optically corrected sensor does not closely match the CIE curve. To remedy this, you need to use a special filter/sensor combination that has correction and calibration parameters for this specific color LED. *Figure 3-2* shows a 652 *nm* red LED's output. If you want to measure the output of a yellow or green LED, use a sensor/filter made for this portion of the spectrum.

The next problem is with unit symbols. You traditionally use the foot-Lambert to measure the light-emitting area of the chip that has a very narrow acceptance angle. This method gives a suitable indication of the light output per unit area, but does not indicate the total output. A chip with a large

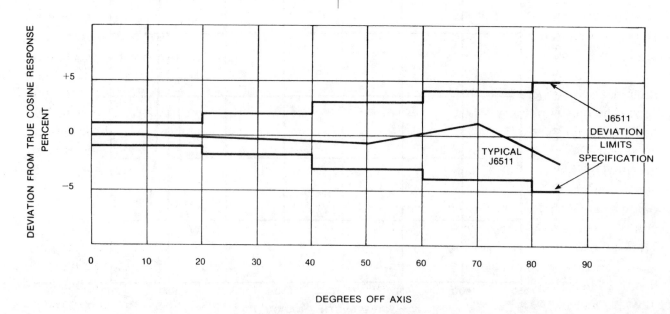

Figure 3-4. *Cosine deviation limits for an photometer sensor (detector). (Courtesy of Tektronix)*

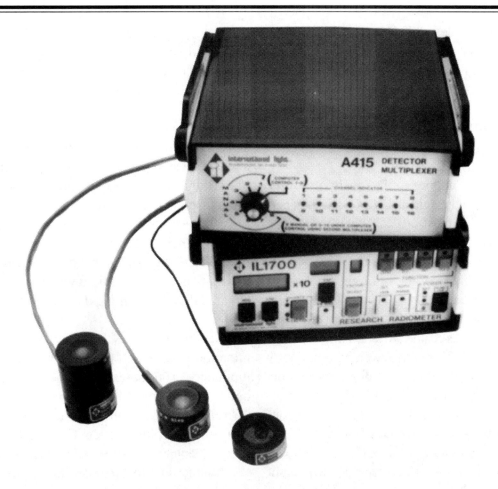

Figure 3-5. *A commercially available radiometer with three optical detectors and a multiplexer from International Light.*
(Courtesy of International Light)

emitting area appears brighter than a smaller emitting area. You need a way of indicating the LED's total useful output.

Illuminance in footcandles more closely meets this need. The only limitation is that you must accurately maintain a known distance to the source because of the Inverse Square Law. A better approach is to use luminous intensity in millicandelas. *Luminous intensity* is a property of the light source independent of distance from the source, although in practice, you also need to control this variable. Since this photometer's sensor is actually 0.2 inches behind the front glass of the filter, you should take this in account. The next problem to overcome is measurement geometry.

LEDs, by design, primarily emit light in the forward direction, although some light exits the LED's sides. Since this "side light" is not useful light, you need an LED adapter. (See *Figure 3-3*.) This consists of baffles which prevent the undesirable light from reaching the sensor. You must also ensure that the LED is on a perpendicular axis with the sensor; however, the holding fixture on the adapter usually secures the LED in the proper orientation.

Practical Considerations

When making low light-level measurements, be certain to ensure against ambient light entering, and use a properly matched filter/sensor combination for the color LED under test. You can also use a short piece of a small diameter plastic rod scribed at the spacing you desire. To guard against reflection and scattering errors, use an ultra-black spray-on paint such as Krylon™ 1602. Lastly, make certain to properly zero the instrument.

Testing Radiometric IR LEDs

Correcting for the Law of Cosines
Cosine correction applies to the sensitivity of the detector receiving the light at angles other than perpendicular, and is most critical when measuring multiple light sources, such as fluorescent lights in an office. At 51° off-axis, the cosine of 51° is .485; therefore, only 48.5% of the full light reaches the detector. As an example, the J6511 has an extremely close cosine correction factor to minimize errors. *Figure 3-4* illustrates cosine deviation limits for this sensor. *Figure 3-5* is a much newer, commercially available radiometer with three radiometric sensors in front of it, from International Light. This radiometer is especially designed to measure photodetector currents over a ten-billion-to-one dynamic range. The device on top is an A415 multiplexer, also from International Light. It provides up to eight

different sensor inputs to the IL1700 radiometer. The multiplexer allows you to select one of these up to eight sensors, and the IL1700 radiometer measures the selected signal.

Reflectance Factor
This closely relates to cosine correction. If you had a white piece of paper which perfectly diffused all of the light striking it (total random reflection), the reflectance would be 1.0 (100% reflectance). *Figure 3-6* illustrates how reflectivity on white paper is greater than on gray paper, and is defined as the ratio of reflected to incident light. Conversely, transmissivity is the ratio of light which passes through a material to the light incident upon its surface.

Practically, this means that 100 footcandles of illumination produces 100 foot-Lamberts of surface luminance (100 footcandles • 1.0 = 100 foot Lamberts). A common, white wall paint typically has a 0.7 reflectance, so you only realize 70 foot-Lamberts of surface luminance. G.E. Bulletin TP-129 lists reflectances of common wall paints; however, you can measure this for yourself. Chapter 17 pre-

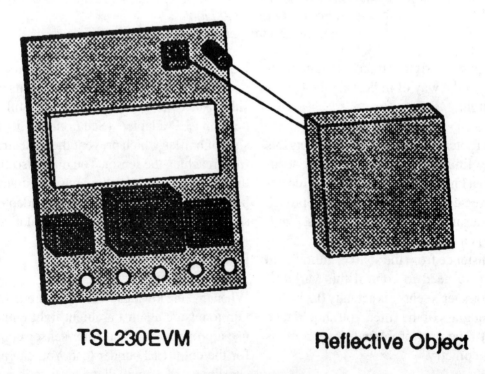

TSL230EVM **Reflective Object**

Figure 3-6. *An illustration of the reflectance principle using different colors of paper. (Courtesy of Texas Instruments)*

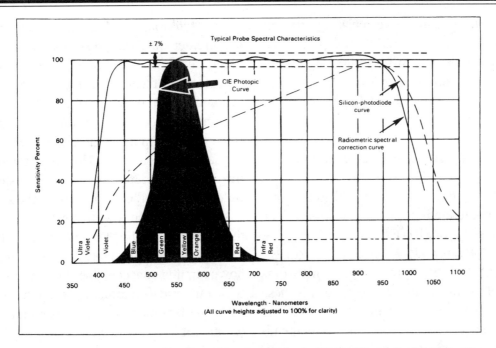

Figure 3-7. *The spectral response of a typical silicon photodiode, with filter. (Courtesy of Tektronix)*

sents an optoelectronic sensor evaluation kit from Texas Instruments with which you can perform a number of experiments, not including the following very simple one. If you buy the kit, place a piece of white paper up to the sensor pair on the board's top right-hand side, then a piece of gray, and finally a piece of black paper. You will obtain progressively smaller light outputs, demonstrating darker paper has less reflectance.

Use the Tektronix J16 and place the J6511 illuminance probe on the surface, then read the illuminance. Use the J6503 probe to measure surface luminance. Divide the luminous reading in foot-Lamberts by the illuminance reading in footcandles, to derive the reflectance factor. It may surprise you, but movie theatre screens and beaded road signs have greater than 100% reflectances. These surfaces purposely reflect light directionally.

Index of Refraction

This describes the relative transmittance of an optical filter with respect to its wavelength. The transmission through the filter is dependent upon the amount of wavelength absorption with the filter material. The index of refraction determines the amount of incident light reflected at the filter/air interface. As an example, a plastic filter with a refraction index of 1.5 reflects 4% of the light, at the interface when the light is normal or perpendicular to the filter.

Using the Proper Spectral Filters and Detectors

Proper color correction refers to the spectral match of a photometer to the CIE curve. Refer back to *Figure 1-2* in Chapter 1, which is a simplified CIE curve. An individual's eyes may deviate considerably from this curve; however, it represents a commonly accepted reference upon which to make photometric measurements. Definition and common international agreements define *footcandles, foot-Lamberts,* and *lux* from detectors having exact color responses to the CIE curve.

A photometer should have a color response as close to the CIE as possible. *Figure 3-7* shows the spectral response curve of a typical silicon photodiode with a glass photo-optic filter, used with the J6511 and J6503 probes. Note how closely the two curves nearly coincide. This close matching ensures accuracy when you measure sources having different colors, such as incandescent, fluorescent, mercury vapor, metal halide, etc. Silicon photodiodes and glass color correction filters provide a better CIE curve match than selenium and cadmium sulfide cells. (See Chapter 5.)

Testing Optical Communications

You can also use the Tektronix J6512 to measure optical fiber attenuation and splice losses. The J6512 irradiance probe has a flat spectral response from 450 to 950 *nm*, and a range of 0.001 to 1999 microwatts/cm². Most optical communications measurements that you will make reside within these parameters; therefore, using the J6512 with the close-coupled detector allows you to interpret readings in just microwatts rather than microwatts/cm². Simply ignore the cm² annunciator on the instrument's front panel.

Figure 3-8 shows a cutaway view of the J6512 with its black plastic cover. You may cut this low-cost cover to fit most optical fibers, connectors, and LEDs. Minimize the distance from the source to the detector to ensure accuracy. *Figure 3-9* shows how reducing the source to detector spacing al-lows you to totally collect light from sources having greater output angles, and from fibers with greater numerical apertures. The numerical aperture refers to the sine of the half angle of the acceptance cone of the light. Some confusion may arise here since, originally, numerical aperture defined the maximum input angle of acceptance, as we have just done here. However, today fiber optics use it as a reference to the measured output field. Typically, the distance from the source to the detector is 0.2 inches with this distance comprising the space behind the glass filter where the detector resides. This is adequate for most applications; however, it is unacceptable, for example, in fibers with numerical apertures greater than 0.74, and LEDs with numerical apertures greater than a 96° output angle.

You measure fiber attenuation by mounting one end of the J6512's black plastic cover (drill a hole in the cover) to a female connector of the type the cable uses. Use a short length of cable (no more than a few inches long) as the reference. Couple an LED to the opposite end of the reference cable, and adjust the current through the LED until you obtain a reading of 1.000. Excessive LED current causes the reading to increase as the LED inter-

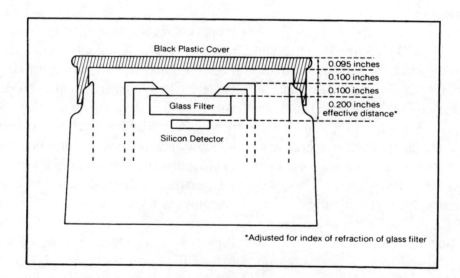

Black Plastic Cover

0.095 inches
0.100 inches
0.100 inches
0.200 inches
effective distance*

Glass Filter

Silicon Detector

*Adjusted for index of refraction of glass filter

Figure 3-8. *A cut away view of a plastic cap you can modify to accept your optical fiber cable or connector.*
(Courtesy of Tektronix)

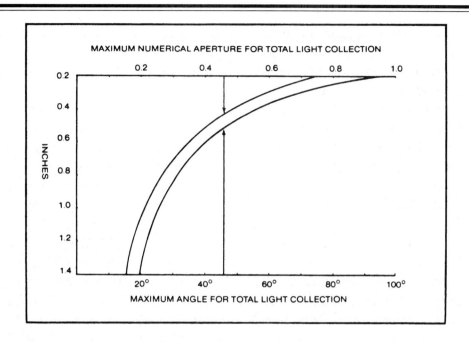

Figure 3-9. *A technique to capture and measure more optical fiber light.*
(Courtesy of Tektronix)

nally warms. Substitute the cable you wish to test in place of the reference cable and note the J16's reading. Use this equation to calculate the dB loss:

$$dB = 10 \log_{10} P_1/P_2$$

P_1 is the sample under test, and P_2 is the reference cable's power reading. You can compute the loss in meters or kilometers by taking the total loss and dividing it by the cable length in meters or kilome-ters, respectively. This same procedure applies to cable splices. When possible, compare readings before and after making a cable splice, and compare the readings to similar lengths of the same type of cable.

Testing the Tester

Figure 3-10 shows how you can, with simple equipment, check the detector's responsivity in the 450-

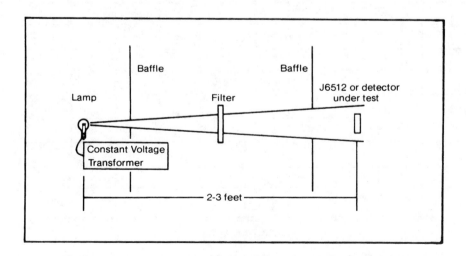

Figure 3-10. *Checking a photodetector's responsivity.*
(Courtesy of Tektronix)

950 *nm* wavelength range. The detector usually expresses responsivity in amps/watts at a wavelength of 850 or 900 *nm*. You measure responsivity at these frequencies with an inference optical filter, and by substitution. First, measure the output current of the detector under test. Then substitute the J6512 in its place, taking care to place it exactly in the same place and at the same angle. Use the irradiance reading from the J16 in microwatts/cm² in this equation to compute responsivity:

$$I/E \cdot A$$

I is the detector output current in microamps. *E* is irradiance of the J16 in microwatts/cm², and *A* is the area in centimeters of the detector being tested.

Testing Photographic Exposure Settings

A camera has numerous variables, such as *f stop*, *shutter speed*, *film speed*, and *scene illumination*, which is what the J16 can very accurately measure. The *f* value of a lens is the ratio of its clear or opened aperture to the focal length. It is assumed that the object is at least 30 times the focal length away. The following equation describes the relationship between these four variables and the constant.

$$t = kN^2/BS$$

k is a constant equal to 3.33, *N* is the *f* stop, *B* is the scene illumination in foot-Lamberts, *S* is the ASA (ISO) rating of your film, and *t* is the exposure time or interval the shutter remains open, expressed in seconds or fractions thereof.

Obviously, you should point the probe directly at the scene you wish to photograph. If you are using a wide-angle lens, you need to point the probe at several angles and objects within the frame of the scene and average these readings. The intentionally small viewing angles of the probes make them insensitive to bright lights or sunlight outside the area of the scene you are measuring.

To determine the correct exposure settings, first select the appropriate graph corresponding to ASA (ISO) film speed. *Figure 3-11* is our representative

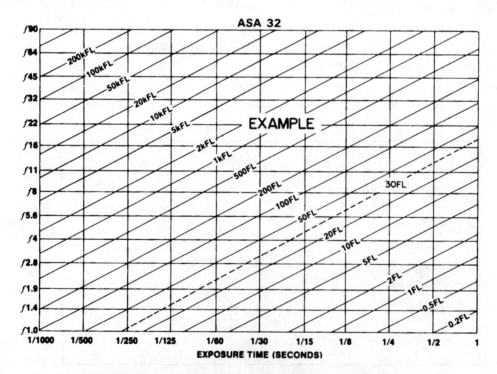

Figure 3-11. *Selecting the right f stop and exposure time for a certain film type and scene illumination. (Courtesy of Tektronix)*

graph for this example, and is an ASA (ISO) 32 low speed, fine grain film suitable for enlargements. Next, find the graph's diagonal line corresponding to the reading you obtained from the J16/J6503 or J16/J6523 combination, and select an *f* stop and exposure time which cross this line. Let's use an example.

If you were using PAN X[tm] film with an ASA (ISO) rating of 32, and you measured a scene illuminance of 30 foot-Lamberts, you would draw a dashed line parallel to and between the 20 and 50 foot-Lambert lines. Referring to *Figure 3-11* again shows that you might select *f8* at 1/4 seconds or *f2.8* at 1/30 seconds as good choices. You may wonder why do this since automatic cameras have built-in light meters — some with multiple silicon photodiodes in a matrix which even integrate or average light. These mass-produced cameras can't correct for photodiode irregularities the way a precise instrument like this does.

You might also make measurements of incident light on the scene, although the accuracy may suffer due to reflectance. That is, a white shirt might have a reflectance factor of 0.9 (90%) and a dark object have 0.1 (10%), with the average around 40 to 50% reflectance. If you measure incident light on the scene in footcandles, you can multiply it by the reflectance factor to get luminance in foot-Lamberts.

Measuring Pulsed-Light in Conjunction with an Oscilloscope

If you wish to measure a pulsed-light source's peak intensity, risetime or waveshape, you may do so with the J16 and any of the Tektronix J6501 to J6505 probes. Once calibrated, you can directly feed these into a scope. The scope has to have at least a 1 mV sensitivity, and you'll need to construct a J16 to BNC adapter. The adapter requires a 15-pin female connector, the connector cover, and a female chassis mount BNC connector. Connect the BNC center pin-to-pin 1, and the BNC shell to pin 2. You'll need a 50 Ω BNC cable and a 50 Ω termination for the operation of the silicon cell.

Calibration and Measurement

You can easily adjust vertical gain to obtain calibration in the same units as read out on the J16:

1. Set up a constant intensity light source, such as an incandescent bulb, to read the light level with the probe connected to the J16. You may see a small amount of 60 Hz ripple on the signal, but you can average this out or use a DC-coupled light source, such as a flashlight.
2. Adjust the probe-to-lamp distance to give a convenient reading on the J16, such as 4.0 µW/cm².
3. Without disturbing the lamp distance or probe position, disconnect the J16 and connect the probe to the scope's vertical input.
4. Use AUTO triggering and set the trace to the center graticule line, using the scope's vertical positioning control with the vertical input GROUND push-button engaged.
5. Release the GROUND push-button, and DC-couple the input.
6. Adjust the VOLTS/DIV switch and control until the trace coincides with a convenient reference point. In our example of 4.0 µW/cm², four divisions would equate to a calibration factor of 1.0 µW/cm².
7. For precautionary measures, recheck the reading on the J16 and the scope's zero trace.

You have now calibrated your scope and can make pulsed source measurements using this factor — providing, of course, that you don't change the VOLTS/DIV setting. You should switch the timebase triggering to INTERNAL unless you are using an external trigger. *Figure 3-12* shows capturing the waveshape of a pulsed-xenon flashlamp light source with a half microsecond risetime.

The xenon flashlamp was invented in the 1930s by Harold E. Edgerton, and provides *microsecond duration* — high intensity pulses of light throughout the ultraviolet to infrared spectrum. This gas-

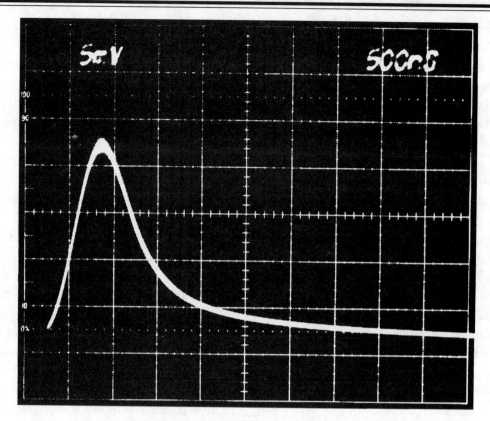

Figure 3-12. *Measuring the output of a xenon light pulse.*
(Courtesy of Tektronix)

filled device (See *Figure 3-13*) converts stored electrical energy into a burst of light by transferring the energy to gas atoms, causing excitation and ultimate ionization. A discharge capacitor usually supplies the electrical energy. These devices exhibit very high peak optical power, but their average optical power is rather low.

Measuring Small Areas of Visible Light

The J6523 1° luminance probe allows you to measure light sources as small as 0.23 inches in diameter over a range of 0.1 to 19,900 foot-Lamberts. Using commonly-available photographic lenses extends this parameter, allowing you to view light sources down to 0.02 inches in diameter. The threaded front housing of the J6523 accepts standard 55 *mm* close-up lenses manufactured by lens companies such as Tiffen, Hoya, and Vivitar. You may also use 48 *mm* lenses by threading onto the J6523 lens instead of its housing. *Figure 3-14* shows the measured light-spot size versus the lens diopter.

This graph represents using two sets of Hoya close-up lens containing +1, +2, and +3 diopter lenses in each.

Close-up lenses use *diopters*. This system rates a lens by its refractive power instead of its focal length. The power of a lens is the reciprocal of its focal length. As an example, assume that a meter is 40 inches long. Then, if a lens has a focal length of 20 inches, that is 0.5 meters or $1/0.5 = 2$ diopters. The + means a positive lens in Plus 1 and Plus 2 lenses; although this is actually +1 and +2 diopters.

Turning the J6523's focus knob fully counterclockwise provides the greatest magnification, but reduces the distance from the object you are measuring to the front of the J6523. With 12 diopters, it is about 3 inches; however, since close-up lenses affect both the viewing and measuring system, you can view the spot area you are measuring through the eyepiece.

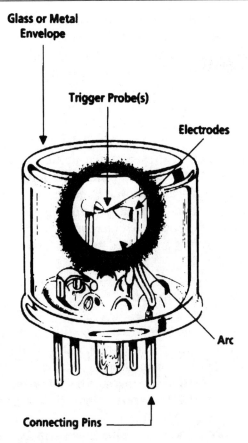

Glass or Metal
Envelope

Trigger Probe(s)

Electrodes

Arc

Connecting Pins

Figure 3-13. *A xenon lamp's structure.*
(Courtesy of EG&G Optoelectronics)

There is an inherent loss of light associated with close-up lenses. The Hoya lenses in this experiment have a 2% loss per lens; the correction factor is therefore 1.02. Using two lens you would have 1.02 x 1.02 = 1.04, three lens about 1.061, etc. The improved coatings on more modern lenses slightly reduces these losses. To determine the loss in a lens, simply measure the center of a uniformly illuminated light source, such as a fluorescent light, both with and without the lens. Be certain to measure the same spot each time while avoiding casting a shadow.

Testing Industrial CRTs in a TV Station

The J16 digital photometer and its various probes allow you to make TV studio measurements, including high resolution slight variances in lighting uniformity. You can measure both luminance and illuminance, and also make color monitor setups. The principal TV studio measurement, though, you'll make with the J16 is monitoring set up for a 6500 K white field. You can determine the red,

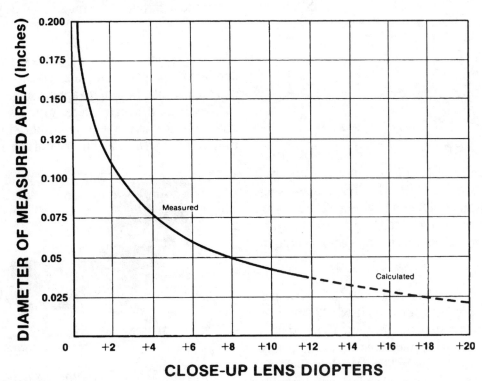

Figure 3-14. *The relationship between a close-up lens' spot size and its diopter rating.*
(Courtesy of Tektronix)

TYPE OF LASER

Media creates Laser beam are as follows ;
Liquid ·····················pigment
Gas ······················He-Ne, Ar and CO_2
Solid ·····················YAG, ruby and glass
Semiconductor ·········GaAs

Laser radiation wavelength zone

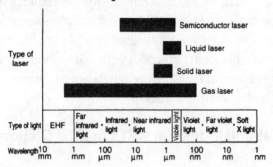

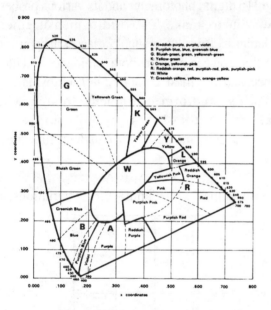

Figure 3-15. A WW rated CRT.
(Courtesy of Tektronix)

green, and blue guns of a color monitor with expected intensities of 10 and 100 IRE. The IRE is a unit symbol from the Institute of Radio Engineers. The until still survives despite the organization's demise long ago due to technological advances. An IRE unit is equal to 7.14 mV.

Since 1945, CRTs in test instruments such as scopes, and consumer products, formerly used a "P" number to describe the persistence of the screen's phosphors. As an example, radar applications use a long persistent phosphor CRT. The WTDS (Worldwide Phosphor Type Designation System) no longer uses *P* numbers, but rather uses one or two letters. These de-

scribe the main characteristics of phosphor screens. The first letter describes the color of the fluorescence and the second letter describes specific differences in the CRT's properties. A white phosphor, previously P4, now is a WW. (See *Figure 3-15*.)

These instruments allow you to quickly judge tracking problems between low and high levels, and correct for them before they become appreciable. Sony Trinitron monitors exhibit great uniformity on manufacturing tolerances; however, earlier monitors had their phosphors vary considerably. Some very early monitors used kinescopes, which require you to initially determine values and record them for monitors on an individual basis.

Monitor Luminance

You can measure luminance in foot-Lamberts on both black-and-white and color monitors with the J6503 luminance probe and the J16 photometer. Operate the monitor at 100 IRE and measure the luminance in the middle of the CRT. Expect a 20 to 50 foot-Lambert reading. You can use the J16 with the J6511 illuminance probe to measure the illumination in footcandles of a grayscale. A *grayscale* is a chart of progressively darker shades of gray placed next to one another. You use this to determine how much contrast your camera delivers.

Place the probe where the chart is with the probe's central axis perpendicular to the chart's surface. You can then adjust lighting to approximately 125 footcandles. Hand held light intensity meters vary by a factor of as much as two to one. Calibrating is simple: place the meters about two feet in front of

Figure 3-16. A commerciallyavailable picoammeter from Keithley Instruments. (Courtesy of Keithley Instruments)

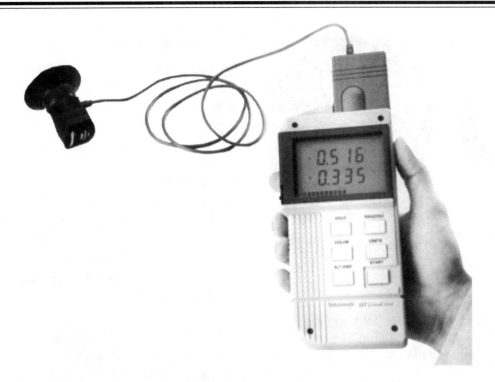

Figure 3-17. *The Tektronix J17 photometer/radiometer/colorimeter, depending on the light sensing probe you select. (Courtesy of Tektronix)*

you and compare their readings against those of the J16/J6511 combination.

Making Actual Measurements

Most manufacturers of photometers make very accurate approximations to the spectral response given by the C.I.E. Standard Observer Luminosity function. This is a compromise of using filters and detector responses with positive deviations at one portion of the light spectrum, offsetting a negative deviation occurring at another portion of the light spectrum. This is excellent for applications attempting to characterize sources with broad spectra (e.g. sunlight, incandescent and fluorescent lamps etc.). However, it is inadequate for narrower light sources such as LEDs (red), which typically occupy the 650 to 660 nanometer spectrum. This requires a photometer calibrated at or near this wavelength.

The spectral response in this wavelength region has a slope of approximately -0.246 dB per nanometer.

Therefore, shifts in this curve may occur due to changes in room temperature, manufacturing variations, and LED drive current. You may use an optical glass filter, such as a Schott-type BG-38 filter, to partially compensate for this. With this combination, the detector is ordinarily much smaller than the LED; therefore, the source-to-sensor distance depends on the diameter of the LED. Assuming a normal-sized LED with a 0.200" diameter, the rule-of-ten tells us to place the detector two inches away. At this distance, a one-millicandela source produces approximately 50 picoamps of current from a PIN photodiode, such as the HP 5082-4207. This may seem very small, and admittedly it is; however, picoammeters can measure these low currents. (See *Figure 3-16*.)

Photosensors may be either vacuum tubes (photomultipliers), or silicon photodiodes. (See Chapter 5.) The problem with photomultipliers (other than being expensive and having bulky power drive requirements) is that they require checking at more

Figure 3-18. *The J1810 (left) and the J1803 (right) light sensors from Tektronix. (Courtesy of Peter A. Keller)*

frequent intervals than silicon photodiodes since they drift with time.

Photometers and Calibrated Photodetectors

Tektronix's Mod J16 photometer with a J6505 LED probe of approximately 1 *cm* in diameter has been correctly calibrated and spectrally sloped for red LED measurements. It has a readout which also directly displays footcandles. This means that if the source-to-sensor distance is exactly one foot, the readout in footcandles is numerically equal to the source intensity in candelas. The one drawback to this system is that the J16, with its J6505 probe, lacks resolution. As an example, at 3.8 inches, the probe is just far enough away to satisfy the "rule-of-ten" guideline for a device with a diameter no greater than 0.38 inches. Otherwise, inaccurate results stem from not following this "rule-of-ten" guideline.

There are many silicon photodiodes on the market. One such photodiode is the Photon Products model VLED-1, which is 1 *cm* in diameter and equipped with a filter providing the proper spectral slope. Calibration usually occurs at 650 *nm*, in microamps per lumen per square centimeter. Also, there is an amplifier designed for converting low photocurrents obtained from low intensity sources directly into voltages with 3-digit resolution readable by a DVM.

Figure 3-19. *The J18 photometer/radiometer/colorimeter used with the J1803 general purpose luminance probe to measure a CRT. (Courtesy of Tektronix)*

	J1803	J1823 or J1823 OPT 01	J1810	J1820	J1811	J1812	J1805	J1806
Capability	Luminance	Luminance	Chromaticity And Luminance	Chromaticity And Luminance	Illuminance	Irradiance	Luminous Intensity	Radiant Intensity
Applications	Displays, Television, Medical	Displays, Television, Lighting	Displays, Television	Displays, Television	Highway Lighting, Office Lighting Transportation	Lase's, IR LEDs, Research	LEDs	Displays, Television
Ranges With J17	0.03 to 1,500,000 cd/m² (Nit), 0.01 to 500,000 footlamberts	0.3 to 30,000 or 3 to 30,000 cd/m² (Nit), 0.1 to 10,000 or 1 to 10,000 footlamberts	Usable with J18 only	0.001 to 0.999 x,y and u',v', 0.3 to 300,000 cd/m² (Nit), 0.1 to 100,000 footlamberts	0.01 to 50,000 lux (lm/m²), 0.001 to 5000 footcandles	0.01 to 20,000 mW/m², 1 nW to 2 mW	0.01 mcd to 99.99 cd	0.001 to 999.9 W/m²/sr
Ranges With J18	0.3 to 300,000 cd/m² (Nit), 0.1 to 100,000 footlamberts	3 to 3000 or 30 to 30,000 cd/m² (Nit), 1 to 10,000 or 10 to 10,000 footlamberts	0.001 to 0.999 x,y and u',v', 0.3 to 1000 cd/m² (Nit), 0.1 to 300 footlamberts	Usable with J17 only	0.01 to 5000 lux (lm/m²), 0.001 to 500 footcandles	0.00' to 2000 mW/m², 0.1 nW to 0.2 mW	0.01 mcd to 10 cd	0.001 to 200 W/m²/sr
Spectral Response	CIE Photopic	CIE Photopic	CIE Tristimulus	CIE Tristimulus	CIE Photopic	Flat	CIE Photopic	Flat
Spectral Accuracy	f_1' = <3% (DIN Class A)	f_1' = <5% (DIN Class B)	f_1' = 3% (DIN Class A)	f_1' = 3% (DIN Class A)	f_1' = 3% (DIN Class A)	±8% (450-950 nm)	f_1' = 3% (DIN Class A)	±8% (450-750 nm)
Accuracy	5% ±1 digit	5% ±1 digit	0.009 at D6500	0.018 at Illum A	5% ±1 digit	5% ±1 digit	5% ±1 digit	5% ±1 digit
Acceptance Angle	8°	1° or 1/3°	16°	16°	180° (Cosine)	96° (approx.)	8°	8°
Other Features	Suction Cup	Optical Sight	Suction Cup	Suction Cup	Level, Cover	Cove'	LED Adapters	
Cable Length	3.5 ft	6 ft	3.5 ft	3.5 ft	6 ft	6 ft	3.5 ft	3.5 ft

Table 3-1. *Probe characteristics versus application and parameters measured for the eight new Tektronix light probes (heads).*
(Courtesy of Tektronix)

EG&G produces models 550 and 450 radiometer/photometers. The 450, with its probe, can read luminous incidance to 2-1/2 digits in either lux or footcandles, and with 3-1/2 digit resolution via the model 550 in either footcandles, lux, or phots. A *phot* is equal to a Lumen/cm², 929 footcandles, or 10,000 lux. These instruments come with a model 550-4 LED filter, and you should use these instead of the photometric filter supplied with the 550-2 probe. This ensures that at the spectral response of red light, there is a proper slope.

More Modern Optoelectronic Measuring Instruments

Tektronik has updated their photometers and radiometers with two new instruments, the J17 TekLumaColor™ and the J18 TekLumaColor™ II. Despite these being newer instruments, however, all the preceding material (including calculations, techniques, precautions and procedures) remain valid. The J17 TekLumaColor™ (See *Figure 3-17*) is for when real-time color isn't critical. It has many useful features, such as:

1. Auto-range.
2. Auto-zero.
3. Hold.
4. Color conversion.
5. Metric conversion to English units.

They have RS-232 ports to allow you to more fully computer automate your testing with storage of all

Figure 3-20. *A typical photodiode, using multi-element glass filters to correct its spectral sensitivity to that of the human eye, used with photometric light probes (heads).*
(Courtesy of Peter A. Keller)

test data. The J18 TekLumaColor™ II allows you to adjust color displays without neglecting your other work since, along with its J1810 chromaticity head (See the left sensor in *Figure 3-18*), it updates and compares readings twice a second. It is calibrated at 6500 K for white light readings. The probe to the right in *Figure 3-18* is the J1803 general purpose luminance probe which you use for CRTs, flat panel displays, and projection screens. *Figure 3-19* shows the J18 and J1803 used with a CRT monitor. *Table 3-1* lists the eight new Tektronix probes per function and capability with either/or the J17 or J18 TekLumaColor™ instruments. *Figure 3-20* is a photoptically corrected silicon photodiode which uses multi-element glass filters to match the spectral sensitivity of the average human eye. These type photodiodes naturally go into photometric, not radiometric sensors.

Chapter 3 Quiz

1. A radiometer measures visible light and a photometer measures non-visible light, T or F.
2. One problem with measuring LEDs is the light energy usually peaks in the red region of the light spectrum, T or F.
3. You can correct this problem by using a special filter/sensor combination, T or F.
4. Reflectance relates a sine factor correction, T or F.
5. (Illumination x reflectance factor) = luminance, T or F.
6. Silicon photodiodes with glass filters provide a better CIE curve match than do:
   A. Selenium cells
   B. Cadmium cells
   C. Both A and B
   D. None of the above
7. Numerical aperture refers to the:
   A. Diameter of a lens
   B. Thickness of a lens
   C. Clarity of a lens
   D. Sine of the half angle of light acceptance
8. You can determine the exposure time of a photo if you know the scene illumination, film speed, and what other variable?
   A. Light meter type
   B. f stop
   C. Type of camera
   D. Film width
9. You can extend the capability of measuring small light sources by using close-up lenses on the photometer, T or F.
10. There is an inherent loss associated with close-up lenses which modern lens coatings have somewhat reduced, T or F.
11. Radar applications use CRTs with phosphors with short persistence, T or F.
12. You use a grayscale in determining the validity of a color, T or F.
13. A spectroradiometer:
    A. Is also called a scanning colorimeter.
    B. Does an immense amount of number crunching internally.
    C. Measures light at each wavelength.
    D. All of the above.
14. The colorimeter's output:
    A. Is a numerically indicated color in one of several color specification systems.
    B. Is more expensive than a spectroradiometer.
    C. Provides better accuracy than a spectroradiometer.
    D. Measures the color of a light source, but not off reflecting objects.
15. A light meter:
    A. Measures only one color.
    B. Does an accurate job of measuring all visible light's intensity
    C. Also accurately follows the spectral response of the color film the photographer uses.
    D. None of the above.

Chapter 4
Visible Light Emitting Sources

Chapter 4
Visible Light Emitting Sources

*"The art of art, the glory of expression, and the sunshine of the **light** of letters, is **simplicity**."*
Walt Whitman (1819-1892), U.S. poet. *Leaves of Grass* Preface (1855).

This chapter covers the drive methods and circuitry for VLS (visible light sources). This is the optoelectronics subject in which you probably have the greatest interest. If you previously viewed LEDs as simple optoelectronic illuminators whose brightness just increased with applied current, this chapter will give you a new appreciation of LEDs and all VLSs! This chapter assumes you are relatively new to these subjects. Thus, there are three "mini-tutorials" on the support electronics which help drive VLSs. These include the operation of:

1. Wave shaping circuits.
2. Op amps.
3. Various memory types, starting with simple diodes comprising an extremely elementary 8-bit ROM.

The chapter covers all the LED VLSs in *Figure 4-1*, which obviously come in a wide range of sizes and configurations.

There are many ways to teach electronics. This chapter emphasizes the laboratory approach, using experiments. Each section therefore characteristically begins with the simplest possible circuits for driving a specific type VLS. These designs are by no means recommended for new products, since this chapter principally stresses basic optoelectronics concepts, not using the latest ICs. Experienced designers might even consider them childishly simple, but their explanations and purposeful orderly progression in technical sophistication definitely help you easily advance to more complex circuits and drive techniques. As an example, the bar graph driving section begins with simple transistor and op amp drive circuits. These demonstrate, in great detail, the difference between a moving dot and a bar graph LED indicator, for example. The bar graph discussion culminates with National Semiconductor's LM3914, an unquestionably viable IC designed for dedicated LED bar graph display driving applications. As a further example, this chapter concludes with a RAM/EPROM dot matrix decoder to drive both types of electrical configurations, characteristic of 5 x 7 dot matrix LED displays. Many cir-

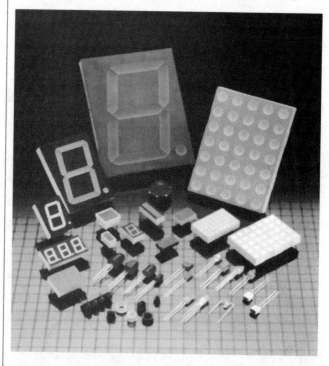

Figure 4-1. *The broad variety of different types and sizes of visible light emitting sources.*

cuits within this chapter use simple logic gate combinations which you realize today with PLAs (Programmable Logic Arrays). These devices allow you to program virtually unlimited combinations of gate and other logic devices. However, before you can implement this technology, you first have to understand the optoelectronics concepts presented here, in as simple a manner as possible. This deliberate approach entails using, among other devices, simple logic gates.

An Overview Sampling of Experiments

"Hands on" experiments start with both single and dual LEDs, and progress to bar graphs, 7-segment, and dot matrix LED displays. A typical experiment is color mixing dual color LEDs with four, and then eight steps, to demonstrate color resolution concepts.

Each variable duty cycle oscillator on the *duty cycle controller* and arrayed LED driver board is uniquely composed of one of the following:

1. Transistors.
2. 555 timer.
3. An op amp.

You'll use these for, among other tasks:

1. Power strobing single LEDs at various duty cycles.
2. Controlling display brightness.
3. Controlling the color shifting in dual color LEDs.
4. Modulating light intensities of a 7-segment LED display, through its decoder/driver.

The chapter's practical portions include explaining how to use formulas, with step-by-step design calculations, especially with the variable duty cycle op amp oscillator. The description evolves over eight equations which describe the driving forces for the circuit's actions. These eight equations not only show you how to calculate this op amp circuit's duty cycles and frequencies, but allow you to calculate values for each of the circuit's 11 discrete passive components.

These three variable duty cycle oscillators assume you do not have access to electronic test signal generators. It is safe to also assume that if you are interested in these experimentation boards, you also like to "tweak" certain key components in a circuit and observe their resulting "causes and effects." This is why you are highly encouraged to first remove the power to a circuit and then "tack" or very lightly solder resistors and capacitors in parallel with each other. This changes both their values, and one or more circuit operating parameters. This often enables you to better understand how a circuit **really** works. Parallel capacitors' effective value equals the sum of the capacitor values. The resultant parallel resistors' value declines as their product is divided by their sum, or:

$$R_{RESULTANT} = R1 \cdot R2/R1 + R2$$

TYPES OF VLSs

LED Lenses

Chapter 2 considered how semiconductor phenomena and fabrication techniques dominantly influenced lens types. This section examines LEDs from a human factor or ergonomics aspect. An LED's visual effect principally differs because of the lens' structure, material and if a lens is even present. **Diffused** LEDs have tiny glass particles in their epoxy plastic lens. These spread (diffuse) emitted visible light, resulting in viewing angles of about +/- 35° off center axis. These LEDs usually protrude through a front panel. **Non-diffused** LEDs have no glass in their epoxy encapsulations and narrower +/- 12° viewing angles. These backlighting (See Chapter 7) LEDs focus on a front panel, which can be as simple as a translucent plastic mylar window.

A **tinted** LED's "off" state has the same color as its "on" state. *Figure 4-2* is a typical LED's construction with an epoxy "magnifying dome lens", which is usually tinted. A **water clear** (non-tinted, non-diffused) LED has the greatest visible light output. It also casts light in the narrowest angle. These LEDs have a "point" effect lens. Viewed off center axis, though, considerable light intensity attenuation occurs. *Figure 4-3* is

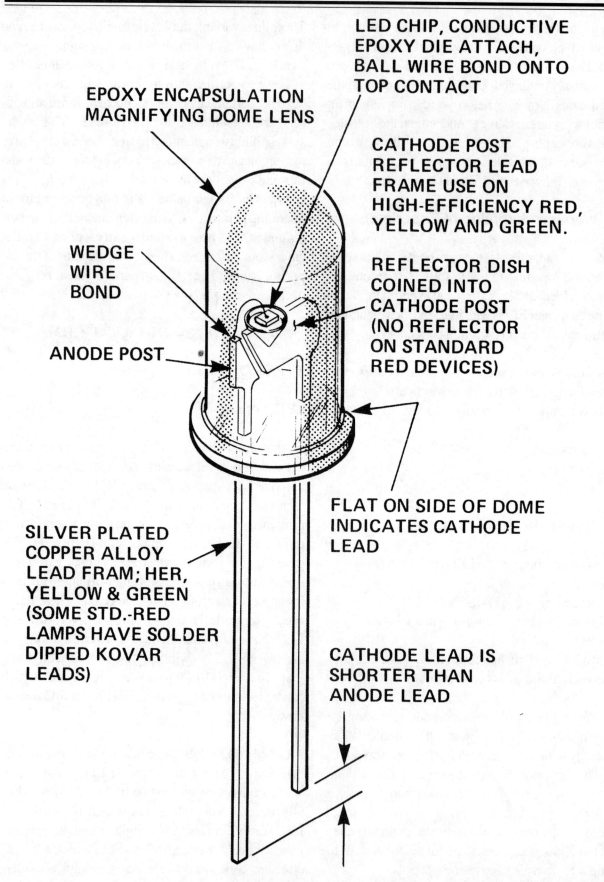

EPOXY ENCAPSULATION
MAGNIFYING DOME LENS

LED CHIP, CONDUCTIVE
EPOXY DIE ATTACH,
BALL WIRE BOND ONTO
TOP CONTACT

CATHODE POST
REFLECTOR LEAD
FRAME USE ON
HIGH-EFFICIENCY RED,
YELLOW AND GREEN.

WEDGE
WIRE
BOND

REFLECTOR DISH
COINED INTO
CATHODE POST
(NO REFLECTOR
ON STANDARD
RED DEVICES)

ANODE POST

SILVER PLATED
COPPER ALLOY
LEAD FRAM; HER,
YELLOW & GREEN
(SOME STD.-RED
LAMPS HAVE SOLDER
DIPPED KOVAR
LEADS)

FLAT ON SIDE OF DOME
INDICATES CATHODE
LEAD

CATHODE LEAD IS
SHORTER THAN
ANODE LEAD

Figure 4-2. *An LED and its domed lens' structure. (Courtesy of Hewlett Packard)*

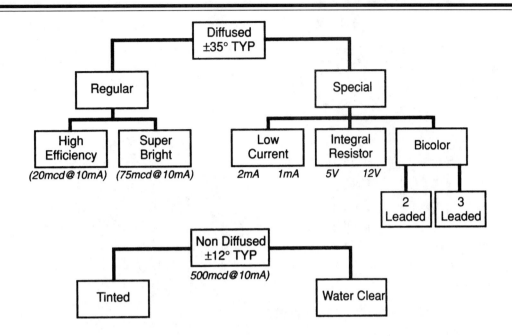

Figure 4-3. *An LED selection chart.*

an LED selection guide which summarizes LEDs by non-diffused and diffused lens types.

LED Sizes and Series Resistors

LEDs have either an internal or external current limiting series resistor. Internal resistor LEDs fuse this resistor onto the semiconductor's substrate. *Figure 4-4* shows subminiature LEDs with internal current limit-

ing resistors. You may recognize them as the "red" nose of a reindeer in an elaborate Christmas card, for example. The three most popular LED sizes are:

1. The 2 mm (T - 3/4).
2. The 3 mm (T - 1).
3. The "jumbo" 5 mm (T - 1 3/4).

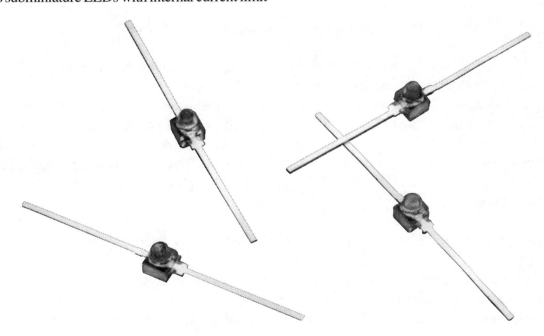

Figure 4-4. *Subminiature LEDs.*

T-1 3/4 or T-1 3/4 LOW DOME LAMPS

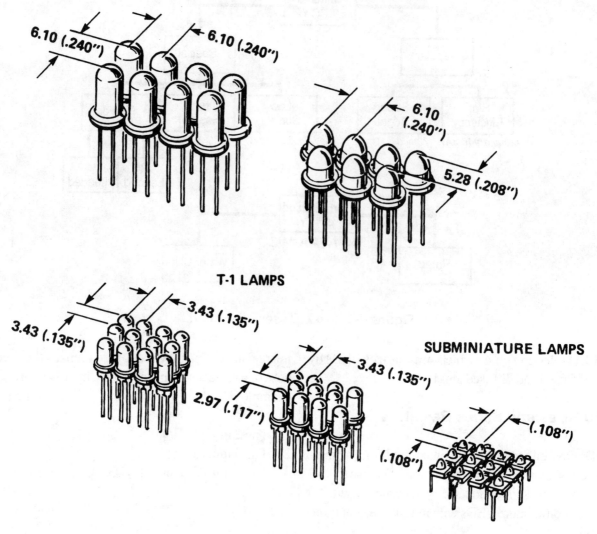

Figure 4-5. Different "T" package sizes of LEDs.

These "T" numbers are atavistic survivors from the older incandescent bulb industry's numbering conventions. *Figure 4-5* shows LEDs in these three sizes, along with a low profile T - 1 3/4 LED.

LED Packages

There are four popular LED packages:

1. The epoxy through-hole LED for PC board mounting.
2. The housed LED delivering light at a right angle from the circuit board. (See *Figure 4-6*) This saves you from manually bending LEDs, but if you have to, follow the technique in *Figure 4-7*.
3. Surface-mountable LEDs, which are mere VLSs without any accompanying optics.
4. LEDs with a prism-type lens structure which bends (diffracts) light parallel to the PC board in rightangle surface-mount compatible packages.

Chapter 8 briefly describes right angle prisms.

Specialty and Evolved LEDs

In the LED's infancy, rectangularly stacked LEDs (See *Figure 4-8*) formed bar graph displays. Single pack-

Figure 4-6. Right angle LEDs.

age bar graphs and ICs to specifically drive them now exist. (See *Figure 4-9*) The rectangular bar graph LEDs' popularity has understandably plummeted. Some bar graphs have up to 101 elements. (See *Figure 4-10*) This further nullifies the practicality of LED

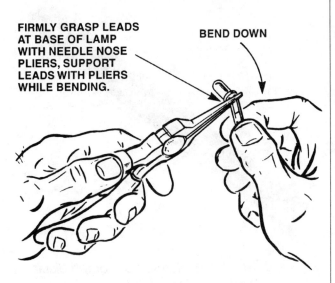

FIRMLY GRASP LEADS
AT BASE OF LAMP
WITH NEEDLE NOSE
PLIERS, SUPPORT
LEADS WITH PLIERS
WHILE BENDING.

BEND DOWN

Figure 4-7. Properly bending an LED to make it a right angle source.

"stacking." Hermetically sealed LEDs for severe environments have metallic bases and sturdy thick plastic encapsulations. (See *Figure 4-11*) These rugged LEDs sometimes require wider PC lands (traces) for their heavier current flow. (See *Figure 4-12*) There are new high efficiency bright (over 100 ft.-Lamberts) LEDs drawing only 2 mA. Chapter 2 concludes with a discussion of a new LED from Hewlett-Packard. To capture a portion of the incandescent bulb market, there are direct LED plug-in replacements for 14 and 24 VDC incandescent lamps, rated at 400,000 hours. These contain multiple physically "clustered" LEDs connected electrically in series. (See *Figure 4-13*) Some LEDs have built in oscillators which flash at about 4 Hz. The Litronix FRL-2000 in 1981 was the first such LED.

Correcting Some LEDs' Viewing Angle Problem

Figure 4-14 shows the ClipLite™, a plastic lens/filter and securing device which solves the point effect of

RECTANGULAR LAMPS

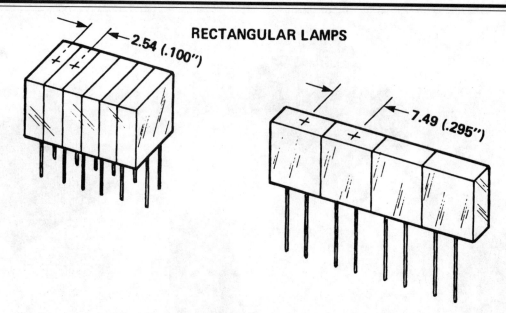

Figure 4-8. Stacking LEDs to form bar graphs.

some LEDs' narrow light radiation pattern. The ClipLite™ allows you to diffuse any kind of LED's light, without suffering significant attenuation of its apparent intensity. The ClipLite™, with its striated (laterally grooved) lines and Fresnel lens, consisting of many small lenses, produces a short focal length while brightly and uniformly dispersing light. (See Chapter 8 for a more detailed description of a Fresnel lens.) It increases apparent power up to 125% and the viewing angle for both diffused and non-diffused LEDs up to 180°.

When installed, these red, yellow, green or clear devices also guard against electrostatic discharge (ESD) to 16 kV. They simply pop into 0.200 inch holes on any flat surface 1/16 to 1/8 inch thick.

LEDs With Silver Leads

Many higher quality LEDs with silver plated leads require extra care in their storage, handling and soldering. Silver chemically reacts with sulfur to form silver sulfide (Ag_2S), or tarnish. This makes soldering diffi-

Figure 4-9. Ten-LED monolithic stackable bar graph light sources. (Courtesy of Hewlett Packard)

Figure 4-10. A 101-element bar graph display. (Courtesy of Hewlett Packard)

Figure 4-11. *Hermetically sealed LEDs. (Courtesy of Hewlett Packard)*

cult. More pronounced tarnish requires more active flux to penetrate and remove it.

Techniques for Handling Silver Plated LEDs

After opening them, place the unused LEDs in tightly sealed plastic bags with petroleum napthelene mothballs to prevent tarnishing. Evaporating napthelene creates a vapor pressure which excludes free air.

Wrapping silver LEDs in tarnish preventing "Silver Saver™" paper also helps.

Solders, Flux and Cleaners for Silver Plated LEDs

Most solders contain approximately 60% tin and 40% lead per Federal Standard QQ-S-571. Two common alternatives are Sn63 eutectic solder and the 2% silver solder, Sn62. Solder flux types, also listed from

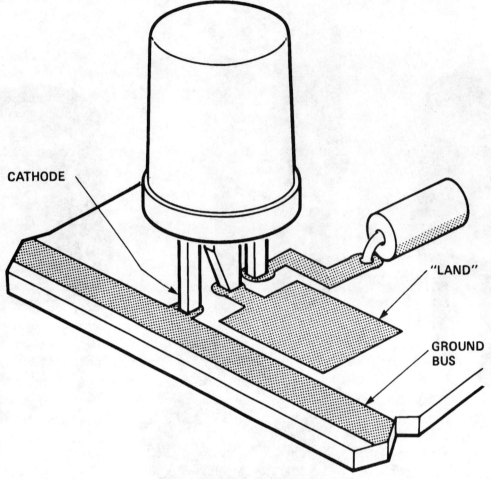

Figure 4-12. *PC board lay out considerations for high current hermetically sealed LEDs*
(Courtesy of Hewlett Packard)

the weakest (R) to the strongest (AC), per Federal Standard QQ-S-571, are:

1. Type R: Non-Activated Rosin Flux.
2. Type RMA: Mildly Activated Rosin Flux.
3. Type RA: Activated Rosin Flux.
4. Type AC: Organic Acid Flux, Water Soluble.

After using RMA or RA fluxes, ionic contaminants exist on the PC board. Use a fluorocarbon F113 and 15% by weight alcohol mixture for removal. You may also wish to investigate other less traditional products which are more environmentally friendly. Using Type AC flux just requires a thorough washing with water.

COMPETING VLS TECHNOLOGIES

Incandescent Lamps

Figure 4-15 shows various smaller incandescent lamp sizes and their bases. (See *Figure 4-16* for some

Figure 4-13. *Physically clustered LEDs connected in series and residing within an incandescent lamp's base socket. (Courtesy of Data Display Products)*

Figure 4-14. *A ClipLite^tm LED lens, holder and securing device. (Courtesy of Visual Communications, Inc.)*

smaller types.) There is an ANSI (American National Standards Institute) three letter code which completely describes both incandescent and halogen filaments. Examples of established incandescent projector lamp filaments are: DMS, BMY, BLC, CEM and DAK. Halogen bulbs include ELH and FCS codes. All incandescent filaments, though, work on the same principle. This is suspending and heating wires (the filament) within a vacuum glass envelope until they glow and emit visible light. This technique has virtually remained unchanged since Thomas Edison discovered

it. A typical bulb is the IDI model 2162 which has a life of 50,000 hours while drawing 80 mA. *Figure 4-17* is a nomograph of three factors controlling light output in candlepower (Lamberts/ft.^{2}). Place a straight-edge across any two incandescent factors to derive the proportions of the missing two. The number of Lumens radiating from an incandescent VLS varies with applied voltage at approximately the 3.6 power. This is especially important in calibrating a photometer with a known output incandescent lamp light standard.

Neon Lamps

These lamps consist of two closely spaced electrodes housed within a gas filled leaded glass envelope. (See *Figure 4-18*) The gas breaks down (ionizes) at a certain voltage, typically 69 volts. After breakdown, the neon lamp runs on its maintaining voltage. This is not the same but close to a neon lamp's excitation voltage, below which gas ionization and current flow ceases. Standard neon lamps typically produce 0.06 lumens per mA, and higher efficiency neon lamps average 0.16 lumens per mA. They last between 10,000 to 50,000 hours which is still just half an LED's life. *Table 4-1* compares the three VLS technologies covered so far. One of a neon lamp's greatest beneficial features though is its ability to absorb voltage surges and transients, without damage!

Figure 4-15. *Various sizes of incandescent lamps. (Courtesy of Data Display Products)*

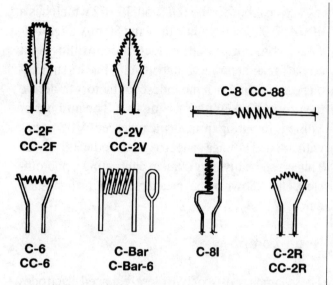

Figure 4-16. *Various incandescent filament types.*

Halogen Lamps

These VLSs greatly resemble tungsten filament lamps, except for their small traces of a halogen (usually bromine) added to the filament. A halogen is any of five elements: fluorine, chlorine, bromine, iodine or astatine within the group VIIA portion of the Periodic Table. In their free states, they exist as diatomic molecules. Adding halogen causes a regenerative cycling of the evaporated tungsten molecules when the bulb's wall reaches 250 °C. Tungsten molecules from the filament combine with halogen vapor which circulate toward the bulb's wall. Its temperature is greater than tungsten's condensing point; therefore, tungsten bromine circulates back to the filament which is about 2,500 °C. This separates these gases. Tungsten deposits on the filament's cooler portion, but bromine continues to circulate.

This lamp's spectral output spans the IR to UV range, making it ideal for use in a spectroradiometer. (See Chapter 3) This intense light output, especially if using a dichroic reflector to focus the light, allows it to drive long fiber optic illumination applications.

THE EXPERIMENTATION PC BOARDS

This chapter has six experimentation boards:

1. The binary code generator (BCG) board.
2. The single and dual LED driver board.
3. The 7-segment driver board.
4. The duty cycle controller and arrayed LED driver board.

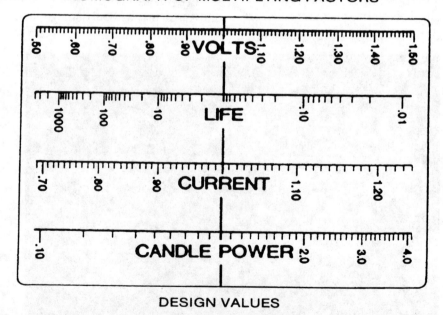

Figure 4-17. *An incandescent lamp nomograph for the relationship of its various salient parameters.*

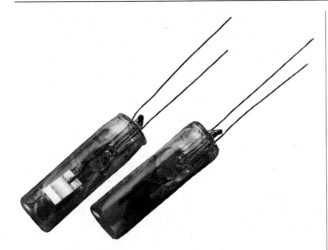

Figure 4-18. *A typical neon lamp. (Courtesy of Gilway Technical Lamp)*

5. The LM3914 10-segment bar graph driver board.
6. The RAM/EPROM 5 x 7 LED dot matrix decoder/driver board.

This chapter presents the first three boards' PC patterns. These each have three drawings, excluding their schematics. The first is the board copper trace foil pattern. These appear just as you would view a finished PC board. Next, there are the component and jumper placement drawings for each of these three boards. Both of these placement drawings have mirror images of the foil pattern as fainter background of the other side as reference images only. Bolder jumper and component placement overlays appear over these faint images. You will have to breadboard the last three boards though, since no PC foil patterns exist.

The BCG Board

There are two prime driving forces in all the experiments. The first is the *duty cycle controller* and *arrayed LED driver board*, which dynamically drives five other boards in a switching manner with variable duty cycle and frequency square waves. The second is the BCG (binary code generator) board's logic level outputs. These statically drive five boards in a non-switching manner. *Figures 4-19, 4-20, 4-21* and *4-22* are the BCG's schematic, foil pattern, jumper placement diagram and component placement drawings

respectively. The non-wiring (component) side's placement drawing also indicates the polarity of components and uses square pads for each LED's cathode and pin 1 of all ICs. The other two experimenter PC boards also follow these "user friendly" practices. You must install jumpers JP1 to JP11 for electrical continuity on this single-sided board, as is true with the other two experimentation board lay outs. Do not install any components or jumpers on any PC board's copper foil pattern side!

There are several paralleled capacitors on each boards' schematic. Refer to the duty cycle controller and arrayed LED driver board in *Figure 4-31* as an example. Note, in the upper right hand corner, there are four such 0.1 mF capacitors, C7, C9, C11 and C14. You might ask yourself, why not just use a capacitor whose value equal or close to their effective capacitance of 0.4 mF? The simple answer is these are bypass capacitors to pass any electrical noise on the power supplies to ground. Also, they are only "bunched" together on the schematic. In reality, these are separate and you strategically place them on each board, solely for bypassing purposes.

The BCG board produces 12 binary logic outputs and indicates, through its LEDs, each line's active or high state. (See *Figure 4-19* again, the schematic) The 12 identical cross-coupled NAND gate pairs are debounced momentary contact (NO) switches, SW1 through SW24. These 12 signals (Binary 1 to Binary 800) form three decades (ones, tens and hundreds) in a binary 1,2,4,8 sequence. These 24 switches therefore create all possible binary patterns of three 4-bit nibbles (half-bytes) or:

1. 1, 2, 4 and 8.
2. 10, 20, 40 and 80.
3. 100, 200, 400 and 800.

For example, a binary 5 has LEDs 1 and 3 (D1 and D3) on while LEDs D2 and D4 remain off.

Transistors Q1 through Q12 sink approximately 8 or 9 mA through their LEDs. This is adequate to brightly illuminate virtually any visible red LED. All 12 binary

	SOLID STATE (LED)	GLOW (NEON)	INCANDESCENT (FILAMENT)
Total light output	Low to moderate* red, amber, green, yellow, blue	Relatively low. Amber-red color, green, blue also available.	Low to high, depending on lamp. Generally white light.
Operating voltage	1.7V and up.	65Vac, 90Vdc for standard brightness types; 95Vac, 135Vdc and up for high-brightness.	1V to 120V (Operation generally not recommended above 28V for non-relampable assemblies).
Operating current	.001 to .100A. Normal operation at .015 to .025A.	.0003 to .005A, depending on lamp.	From .015A up, depending on lamp.
Ballast required (current-limiting resistor)	Yes. Rectifier diode also required for ac service.	Yes. Normally built into housing or installed in one lead for non-relampable assemblies.	No. (Voltage-dropping resistor sometimes used).
Life	100,000 + hours	Minimum 5000 hours. up to 50,000 + depending on current.	Depends on lamp life rating and conditions of use.
Resistance to vibration, shock, voltage transients	Very high.	Very high.	Low to moderate, depending on lamp, type of shock.
Cost	Low to moderate,	Low.	Low to high, depending on lamp.
Typical applications	Battery operated devices with integrated circuits, low current applications.	Line-voltage operated appliances, instruments.	Very broad range.

*Recent improvements in material technology has increased brightness levels significantly.

Table 4-1. *Comparing three VLS technologies. (Courtesy of Industrial Devices, Inc.)*

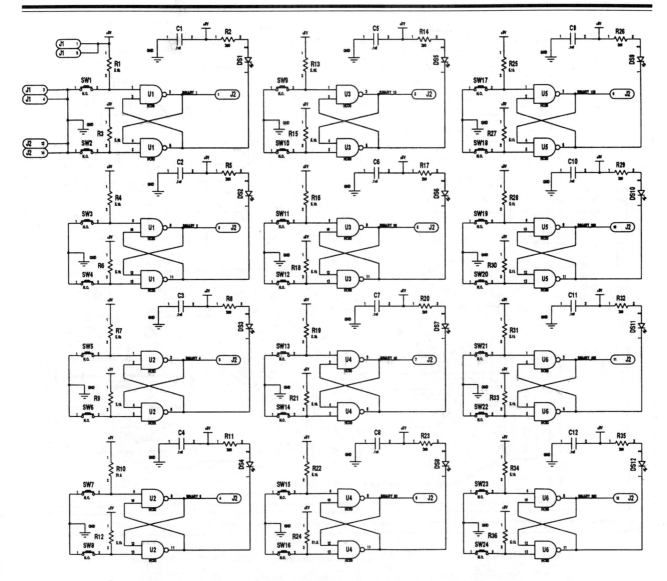

Figure 4-19. The BCG (Binary Code Generator) schematic.

signals originate from the board's edge. (See *Figure 4-22* again) These signals connect to other boards through a flat rigid ribbon cable type connector.

The Single and Dual LED Driver Board

Figure 4-23 is the single and dual LED driver board schematic and has two transistors alternately turning on a dual color LED. There are four different transistor drives demonstrating how to turn on a single LED. Two gates alternately turn on two different LEDs. There is a J-K flip-flop to turn on another dual color LED, but this circuit ensures a perfectly symmetrical square wave. This board has a slow and a fast square wave oscillator, and means to drive this J-K flip-flop and

concludes with two 74HCT151 data selectors mixing a dual color LED's colors in eight steps. *Figures 4-24* to *4-26* are this board's PC foil pattern, its jumper placement and components placement drawings, respectively.

The 7-Segment LED Driver Board

Figure 4-27 is the schematic of this board. This board "fixes" several annoying 7-segment display characteristics. These include incompletely forming the numerals "6" and "9" and displaying the entire hex set to decimal 15 or hex F. *Figures 4-28* to *4-30* are this board's PC foil pattern, its jumper placement and components placement drawings respectively.

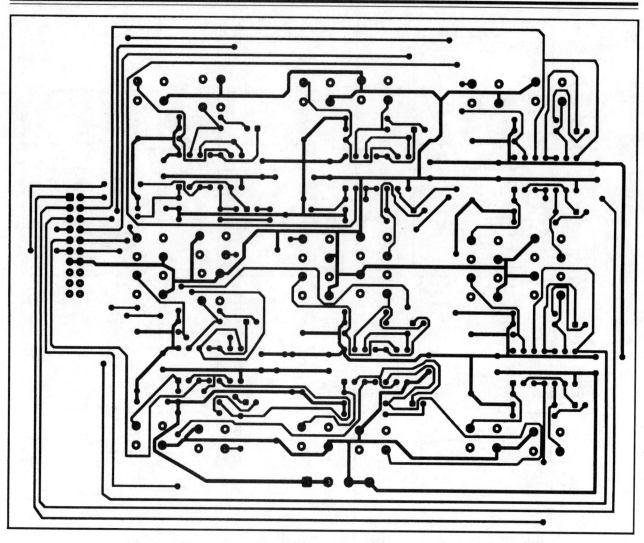

Figure 4-20. *The BCG (Binary Code Generator) board's PC foil pattern.*

Design Basics

As mentioned, the duty cycle controller and arrayed LED driver board (See *Figure 4-31*) has three types of oscillators, all based on a different active device, and all with difference duty cycles, and even different frequencies, is you elect to design them that way. A duty cycle is the ratio of a square wave's ON time to its OFF time. The 10% duty cycle oscillator is therefore on for 10% of the time and off the remaining 90%. Note all components' values in the oscillators and power strobing circuit are purposely blank. These are left as exercises for you to solve, helped by the following oscillator primer based on duty cycle and wave shaping theory, and your design analyses.

Wave Shaping and Duty Cycle Theory

An oscillator's design differs from that of an amplifier because its input and output are in phase. An amplifier's input and output are 180° out of phase. An amplifier also provides gain, but to sustain oscillation, an oscillator just barely maintains unity gain. The three variable duty cycle oscillators operate in a digital, not an analog or linear fashion. Ideally, a digital or square wave oscillator quickly drives one of the oscillator's transistors (or other active devices) into saturation while the circuit's other transistor cuts off.

A minimum interval saturation-to-cut off swing, and back again, ensures steeper (nearly 90°) square wave edges. The longer these transitions take, the greater

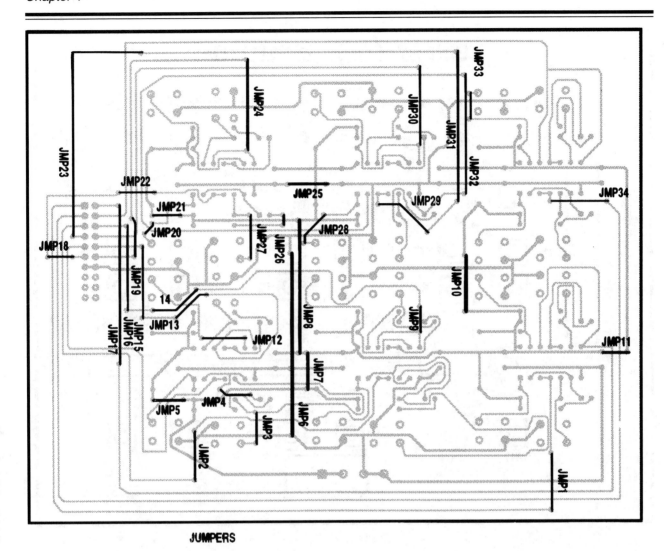

Figure 4-21. *The BCG (Binary Code Generator) jumper placement drawing.*

the square wave's rise and fall times. A rise time is the time it takes a square wave's rising edge to change from 10% to 90% of maximum amplitude. A fall time is the same, only measured on the square wave's falling edge from 90% to 10%. A slow switching transistor idly dwelling in its linear region, between saturation and cut off, produces slow rise and fall times.

Triangular wave shape oscillator designs purposely dwell longer in the linear region. A triangle wave oscillator's rising and falling ramps result from a gradual slow **ramping** transition in the transistor (or other active device you elect to use). We use a triangular shaped wave form in the "dithering" LM3914 circuit but realize it with an op amp and a series RC combination on the op amp's output. Its wave shape is very

sine wave like though at its peaks of 90° and 270°. A sawtooth waveshape results from a slow transition in one switching transistor and a fast switching transition in the other. This causes a wave shape which slowly ramps up and then abruptly returns to zero.

A sine wave oscillator constantly dwells or operates in the linear region, experiencing **no** abrupt switching states. Such an oscillator usually involves both positive and negative feedback to provide a zero phase at just one frequency. This is its output frequency. *Table 4-2* is a comparison of these four wave shapes referenced to a symmetrical square wave, and assumes all the other wave shapes are symmetrical.

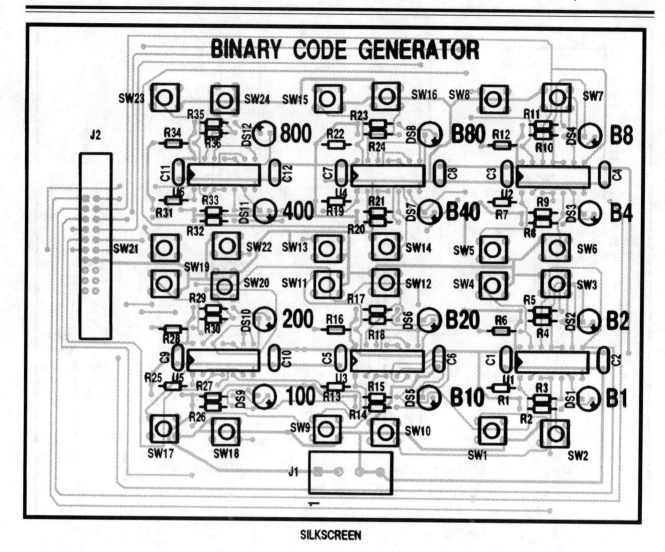

Figure 4-22. *The BCG (Binary Code Generator) board's component placement drawing.*

The Transistor-Based Astable Oscillator

This design is the **simplest** possible free running or astable oscillator, see the upper left circuit in *Figure 4-31*. The values of the two resistor and capacitor combinations determine this astable oscillator's frequency, period and duty cycle. The following equation defines a period, T, and its relationship to frequency, *f*. T is one whole cycle, or period:

$$T = 1/f$$

The next equation defines a period's two constituent parts, which are the sum of the ON and OFF times:

$$T = t_1 (t_{ON}) + t_2 (t_{OFF})$$

Both transistor base resistors and charging capacitors are equal, T = 1.38 RC with R, the base resistors (R2 and R5) in ohms and C, the collector-to-emitter capacitors (C1 and C2) in farads. Temporarily ignore the pot, R1, which later proves to be very important.

Since capacitor tolerances are typically 10% or greater, you may have difficulty if you desire an exactly symmetrical square wave. To guarantee this, consider doubling the square wave's frequency and apply it to a flip-flop. This divides the square wave's frequency in half, but with perfect symmetry in its output.

If you use different (not matching) values for the RC product time constant (τ) determining components (See *Figure 4-31* again), you will have unequal tran-

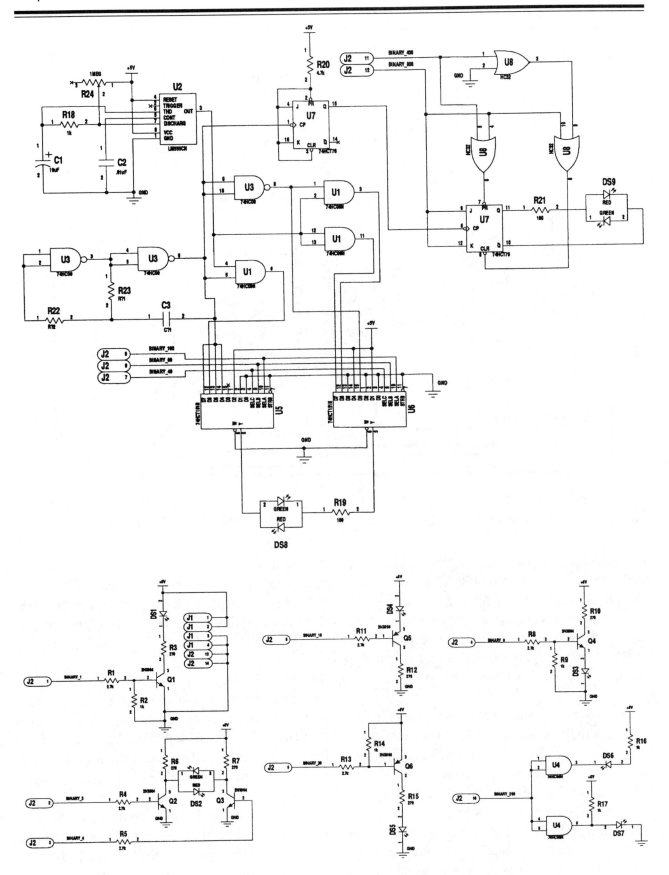

Figure 4-23. *The Single and Dual LED driver schematic.*

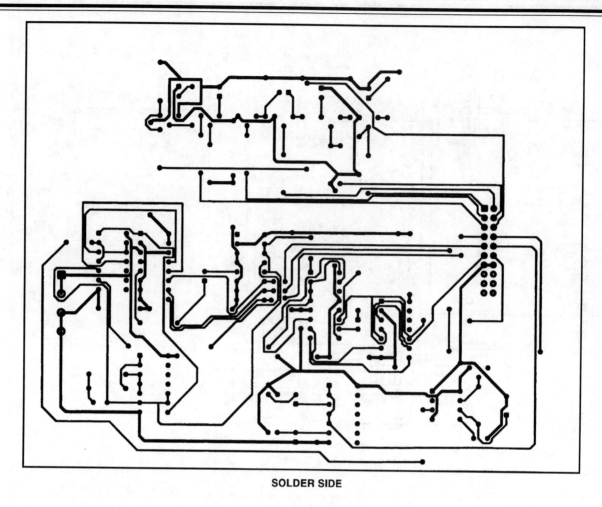

SOLDER SIDE

***Figure 4-24**. The Single and Dual LED driver board's PC foil pattern.*

sistor conductions and not a 50% duty cycle. This the basis of your asymmetrical (not 50% duty cycle) square wave oscillator. In any free running or astable oscillator, one transistor charges which shuts off after charging the other transistor, through its base capacitor. This back and forth charging and discharging recirculating action creates square wave oscillations.

Design this circuit to be a symmetrical astable oscillator, that's right, symmetrical! Then notice the previously ignored pot, R1, and make its value at least seven or eight times that of either transistor's base resistor. This intentionally makes the two RC pair products (τ) unequal, which accomplishes our goal of an unsymmetrical waveform (non-50% duty cycle). This provides a wide range of control over the duty cycle. Incidentally, what if, for example, you do have a 10% duty cycle at the collector of transistor Q1. What do

you think the duty cycle will be on the collector of Q2? What is Q2's ON time or duty cycle? It should be 90% since the previous equation defines a period, T, as composed of **both** the ON and OFF times. It should be 100% — 10% or 90%.

Two AC cross-coupled transistors from one's collector to the other's base, and vice versa, was the way engineers used to design astable squarewave oscillators. But occasionally, there is still a need for this type design in modern applications. *Figure 4-32* is such a circuit, which allows you to use a single alkaline, NiMH or Ni-Cd battery in the 1.25 to 1.5 VDC range to drive circuits requiring 1.8 VDC. The new low supply voltage op amps from National Semiconductor, such as their LMC6582 and LMC6681 families are examples. These op amps have a gain bandwidth product in excess of a million.

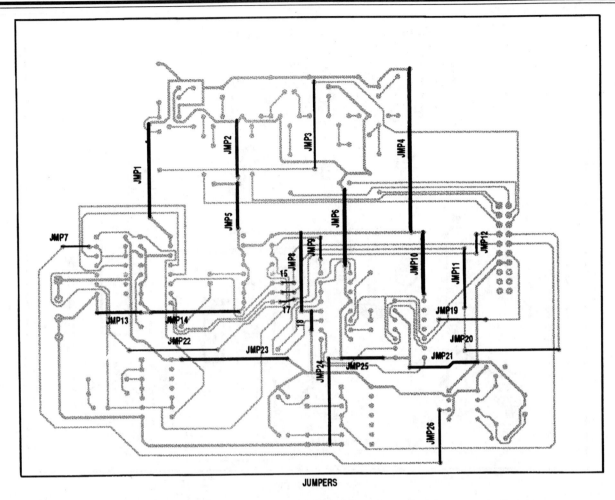

Figure 4-25. *The Single and Dual LED driver jumper placement drawing.*

Your first instincts might be to design a DC-to-DC converter without an electrically noise inductor. However, there are no affordable inductorless ICs at present which do this for less than $2.50, even priced in quantities of 1,000. *Figure 4-32* solves this with transistors Q3 and Q4 forming an astable oscillator with a symmetrical (50% duty cycle) output since its RC time constant (t) determining pairs, R_1C_2 and R_2C_1, are equal. They also determine the circuit's approximate 50 kHz output frequency. This circuit's active complementary load is Q1 and Q2. Transistors Q1 and Q2 conduct 50% of the time on each alternate half cycle and are exactly in step with their astable oscillator's drivers, Q3 and Q4. This circuit operates from as low as 1.0 VDC. The green LED is optional, but provides a simple low voltage parallel (shunt) regulator and, a power ON indication. The table shows the circuit's performance with and without the

LED and its 1.9 volt forward voltage drop. It also shows how the load resistor (with the LED in place) influences the output voltage and the ripple's magnitude. This circuit can drive the stereo base-expander in *Figure 4-32.*

The 555 Timer Oscillator

The 555 timer has been around since Signetics introduced it in 1972, and it is still a popular IC. This is a far simpler circuit to analyze, and *Figure 4-33* is the simplest possible 555 timer astable or free running square wave oscillator. Again, as in all oscillators, T = t_1 (the on time) + t_2 (the off time). The timing resistor and capacitor, R_t and C_t, solely determine its period, T, and frequency, or 1/T. Theoretically, it has a symmetrical output; however, a low, +5 VDC supply (for example) on pins 4 and 8, causes some asymmetry as

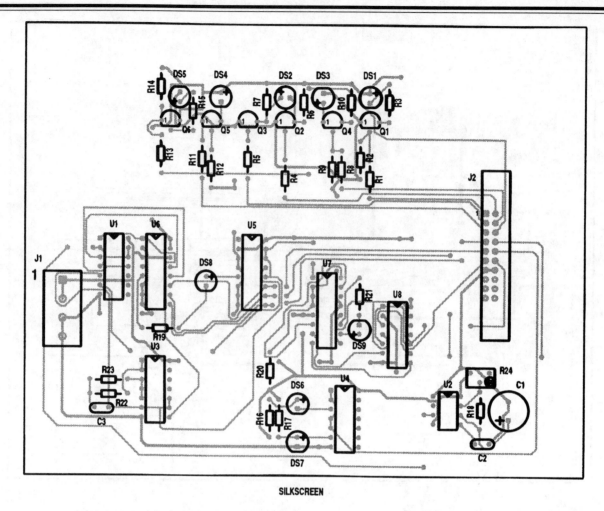

Figure 4-26. *The Single and Dual LED driver board's component placement drawing.*

the square wave output approaches the 555's high saturation state, which is nearly 2 VDC below the +5 VDC supply. This, and its sheer power consumption, are why a 555 timer would never be used in a modern design operating at or near 3.3 volts. *Figure 4-34* is virtually an identical circuit with all its component arrangements and pin outs; however, it uses a 7555 low power consuming CMOS version of the older 555 timer. This takes advantage of this IC's swings from its supply voltage to ground. This wider threshold of output voltage swings eliminated the 555's tendency toward asymmetry at lower supply voltages. The following equation is the period, T, of this circuit. That is why our 555 timer astable oscillator uses +12 VDC as it supply voltage.

$$T = 1.39 \, R_t \cdot C_t$$

The variable duty cycle oscillator in *Figure 4-31* uses two transistors, Q5 and Q6, for a 5% to 95% duty cycle **without** changing the period, T. During the square wave's OFF time, Q5 is on and Q6 is off, just like our previous two transistor circuit in *Figure 4-31*, composed of Q1 and Q2. This disconnects the V_{DD} as the timing capacitor, C3, discharges into the 555's discharge pin (7). During the square wave's ON time, Q6 reconnects V_{DD} to charge C3. The pot, R20, adjusts the charging resistance, composed of the total resistance from Q6's collector to the 555's pins 2 and 6 (going through the pot's wiper). This proportionately increases the ON time. Decreasing R20 has an equal but naturally opposite effect of increasing the OFF time, at the ON time's expense, while maintaining your predetermined period, T. Resistor R19 protects Q6 against discharge and charge current surges.

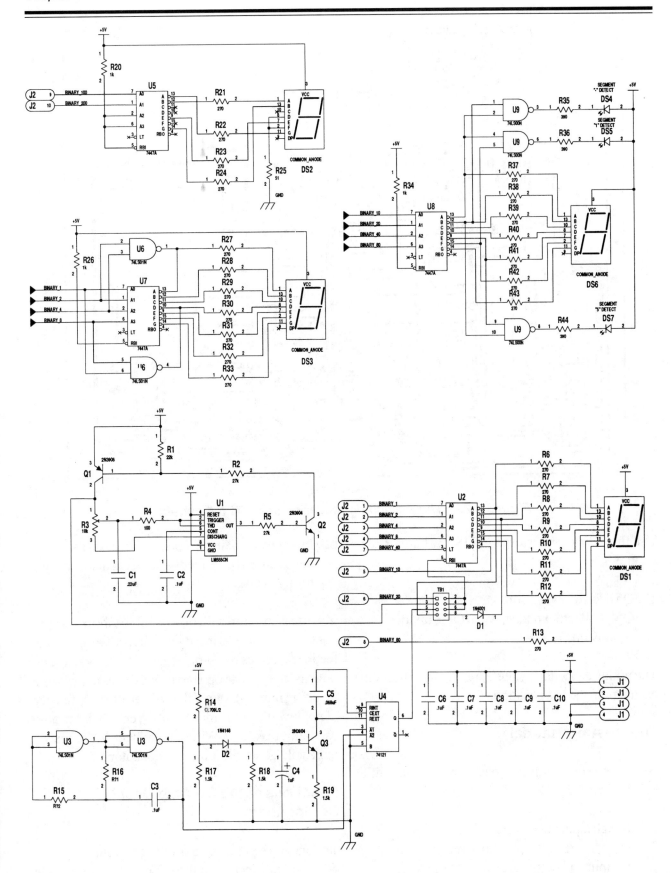

Figure 4-27. *The 7-Segment driver schematic.*

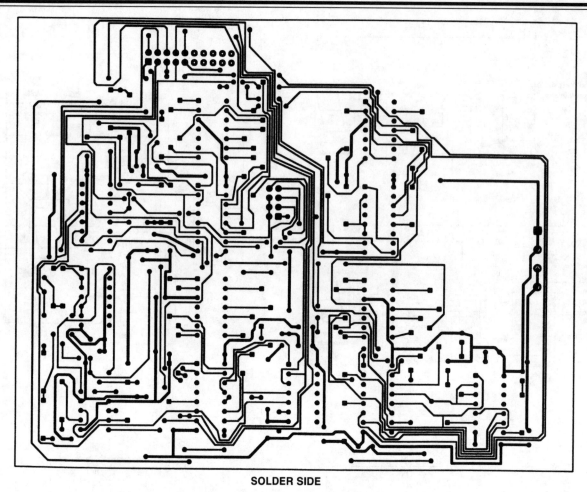

SOLDER SIDE

Figure 4-28. *The 7-Segment driver board's PC foil pattern.*

Use three 20 kΩ resistors for R18, R21 and R22 to ensure:

1. Q5 decisively turns on.
2. Q5's collector voltage, when OFF, divides equally on Q6's base.

The timing capacitor's value approximates that of the capacitor in the last equation.

The Op Amp Tutorial

An op amp very nearly approaches an ideal amplifier in at least five aspects:

1. Its high input impedance.
2. Its perfect feedback control conformance.
3. Its input and output are 180° out of phase.
4. Its minuscule input current.
5. Its ability to maintain equal voltages on its inputs, caused by its feedback mechanism always trying to do so.

You use an op amp with or without feedback components. An op amp is input level sensitive and, using feedback, drives its output to equalize the voltage on both its inputs. The op amp usually tries to make the inverting input (-) follow the voltage on the reference or noninverting (+) input, or vice-versa in the op amp oscillator to follow. Without feedback, op amps are mere voltage sensitive level comparators.

Functions of the Op Amp Oscillator's Components

The op amp square wave oscillator circuit in *Figure 4-31* creates selectable pulse widths, or is a duty cycle controller, as our two preceding circuits are. It pur-

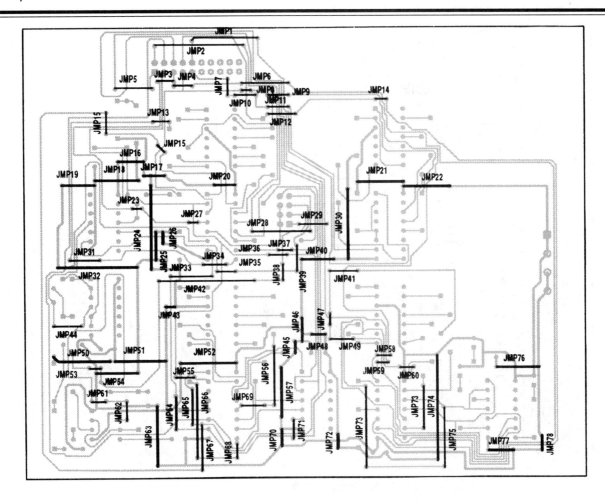

Figure 4-29. *The 7-Segment driver jumper placement drawing.*

posely overdrives its output to form square waves. The U3 op amp circuit uses both positive and negative feedback and is composed of three smaller integral circuits, which are:

1. An integrator formed by C5 and R36 or R37, depending on which small signal diode (D1 or D2) is conducting.
2. A voltage level sensitive op amp comparator, which an op amp's inherent nature provides.
3. A latch which uses the positive feedback network (going to the (+) input), composed of resistors R35, R38 and pot R39 (to its wiper).

Resistor R34 controls the current through zener diodes D3 and D4. The small signal diodes, D1 and D2, conduct during the ON and OFF times which comprise the period, T. The two zeners, D3 and D4,

set the output waveform's positive and negative limits. The pot, R39, varies the duty cycle, covered in detail later.

Circuit Operation

Let's call the positive feedback ratio of R35/[R38 and R39, to its wiper], β, and assume the output has just switched to zener diode D4's positive voltage. Almost simultaneously, the op amp's (-) input is at $-\beta V_{D3}$'s voltage (the OFF state) and the op amp's (+) input has just switched to the voltage of $+\beta V_{D4}$ (the ON state). The feedback, β, through R35, causes the (+) input voltage to be more positive than the op amp's (-) input. The op amp temporarily remains locked in this unnatural condition (its latching function) of unequal input voltages. These "unnatural" states are its ON and OFF times. In fact, the inputs are almost never equal, but are constantly correcting this inequality by

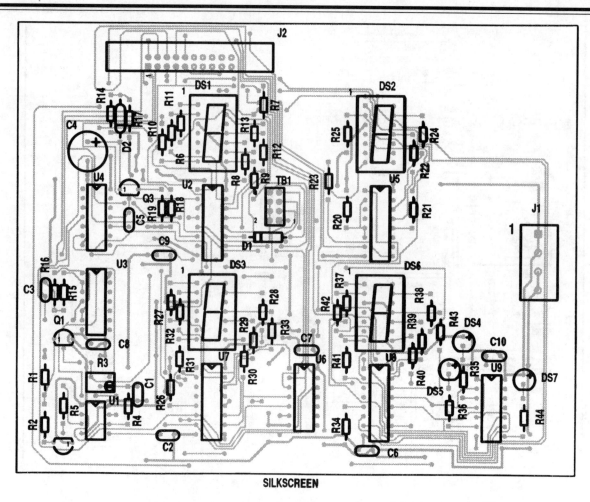

Figure 4-30. *The 7-Segment driver board's component placement drawing.*

its output inducing current flow through D1 and R37, or D2 and R36. As the op amp square wave switches from ON to OFF, or back again, its inputs momentarily equal each other, but just in passing, literally.

To be precise, add 0.65 volts to the zener voltage output voltage since one of the zeners is forward biased, and is not breaking down, so it drops this much forward voltage. V_D in the equations represents this forward biased diode's drop of 0.65 volts. This causes output switching. This charges capacitor C5, at the inverting (-) input node, as the discharging following equation describes:

$$v_1 = (\beta V_{D3} + V_{D4} - V_D) [1 - \varepsilon^{-t/(R37\ C5)}] - \beta V_{D3}$$

When this voltage attains a level slightly above +βD3's (the ON time voltage), the op amp switches states

and the output goes to -βD4's voltage (the OFF time). The following equation describes this charging of capacitor C5, with $+V_{D4}$ = the ON period and $-V_{D3}$ = the OFF period:

$$v_1 = (V_{D3} + \beta V_{D4} - V_D) [\varepsilon^{-t/(R36\ C5)}] - (V_{D3} - V_D)$$

The op amp constantly tries to balance or have equal inputs as the two previous equations describe. Adding these times and taking their reciprocal equals the square wave's frequency:

$$f = 1/[T_1 + T_2] = 1/[T_{ON} + T_{OFF}]$$
$$= (R36 + R37)/R37$$

The next equation describes their duty cycles, which is what we primarily seek. The duty cycle of 60% re-

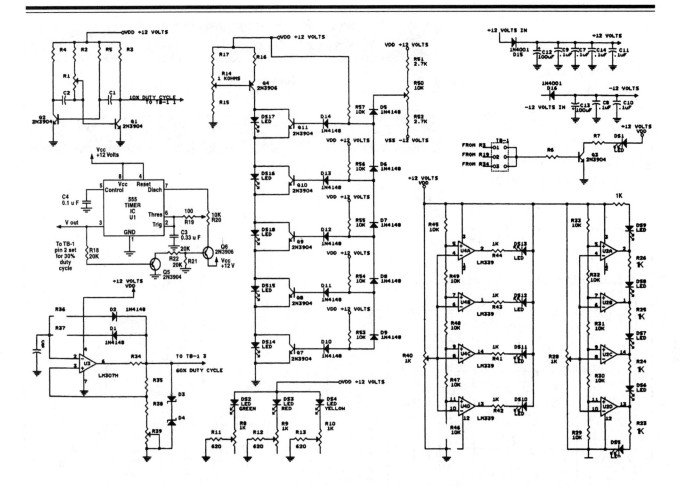

Figure 4-31. *The Duty Cycle Controller & Arrayed LED Driver Board schematic.*

quires R38 + R39 (to its wiper)/[R38 + R39 (to its wiper) + R35] to be 0.6:

$$\textbf{Duty Cycle (D)} = \textbf{R37/[R36 + R37]}$$
$$= \textbf{T}_{ON}/[\textbf{T}_{ON} + \textbf{T}_{OFF}]$$

Design Steps and Considerations

In your design, start with a 60% duty cycle goal and make the [ON/ON + OFF times] = 0.6. There are maximum values for the feedback resistors R36 and R37 which control capacitor C5's charging current. Determine these maximum values of resistors R36 and R37, using the next two equations. These are directly proportional to, and again control, the ON and OFF times. Use a 30% maximum value for R36 and R37 to bias both equal value zeners well into their breakdown regions:

Max Value for R37 $\ll V_{D4}/I_{MAX}$

Max Value for R36 $\ll V_{D3}/I_{MAX}$

Determine the ON and OFF times by the next two equations. This will first require you making a reasonable selection of C5, suggested values are 0.1 µF or less. The higher frequency you desire, the smaller the values of C5 and R36 and/or R37 will be, again, depending on which diode, D1 or D2, is conducting:

ON Time Interval in seconds
$$\textbf{T}_{ON} = \textbf{R37 C5 ln } [\textbf{V}_{D4} + \beta\textbf{V}_{D3}$$
$$- \textbf{V}_D/(\textbf{V}_{D4} (1 - \beta) - \textbf{V}_D)]$$

OFF Time Interval in seconds
$$\textbf{T}_{OFF} = \textbf{R36 C5 ln } [\textbf{V}_{D3} + \beta\textbf{V}_{D4}$$
$$- \textbf{V}_D/(\textbf{V}_{D3} (1 - \beta) - \textbf{V}_D)]$$

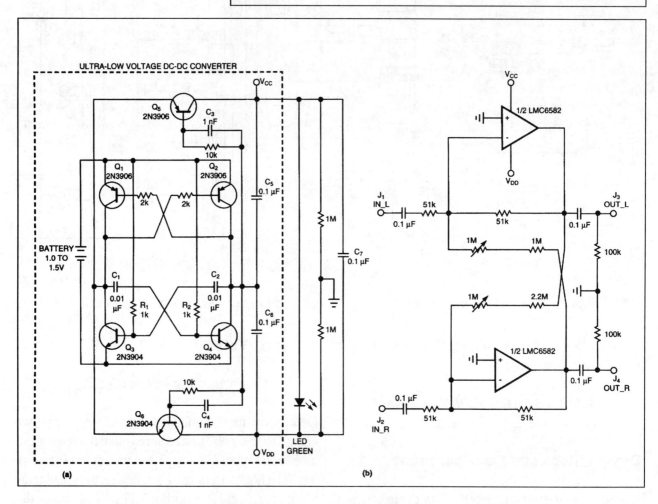

TABLE 1—V_OUT VS V_IN (WITH LED, NO LOAD)

V_{IN} (V)	0.7	0.8	0.9	1.0	1.1	1.2	1.3	1.4	1.5	1.6
V_{OUT} (V)	1.1	1.37	1.6	1.75	1.81	1.85	1.89	1.91	1.93	1.93

TABLE 2—LOAD CAPABILITY AND RIPPLE (WITH LED)

R_{LOAD} (kΩ)	0.5	1	2	3	5	10
V_{OUT} (V)	1.74	1.84	1.87	1.88	1.89	1.89
Ripple (mV p-p)	5	4	4	4	3	2.5

Figure 4-32. An inductorless DC-to-DC converter accommodating a single cell of various battery technologies. (Courtesy of EDN Magazine & Chaners Publishing)

Pot R39 varies the duty cycle. Selecting a large value for R39, compared to R35 and R38, provides a wide range of duty cycles, but coarse adjustment resolution. Conversely, the opposite produces a smaller duty cycle range. Using a smaller value 10-turn pot, compared to R35 and R38, produces a smaller range **and** very fine adjustment resolution.

Don't Be Blinded By the Light

The previous equations may initially appear formidable, but they are not! Very gingerly review them and ask yourself what they really describe while referring to each op amp oscillator's component. Then, see how the equations, which appear somewhat similar to you, differ. For example, the fifth and sixth equations differ mainly because they describe the series RC charging and discharging of capacitor C5, from current flowing

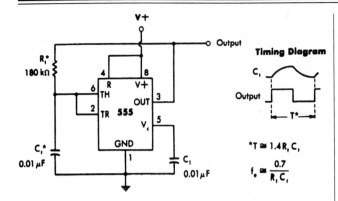

Figure 4-33. *The simplest possible 555 timer astable oscillator.*

through resistors R37 and R36, respectively. They also differ where β is multiplied by one or the other zener diode. In the seventh equation, the frequency **f** = R36 + R37/R37. Can you derive this from the eighth equation, which describes duty cycle? First realize that duty cycle, **D**, and frequency, **f**, are related as reciprocal quantities, or f = 1/D and D = 1/f. Therefore, if the duty cycle, D = R37/R36 + R37, its frequency equals the reciprocal of that, or R36 + R37/R37. You merely flip over this resistor fraction involving R36 and R37. Analyze these equations in steps, and if you are rusty in this area, review the equations describing the charging and discharging of a series RC combination.

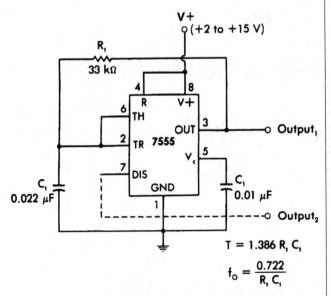

Figure 4-34. *A CMOS 7555 version of the older 555 timer that eliminates the tendency toward asymmetry which, when changed, does not affect its period.*

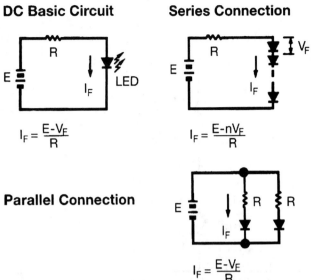

Figure 4-35. *Calculating LED current flow and series limiting resistor values in single, multiple and parallel LEDs.*

EXPERIMENTING WITH LEDs

Power Strobing LEDs

Do not summarily gloss over this subject due to its apparent simplicity. This is a mandatory technique to operate all 7-segment multiplexed LED displays and **every** dot matrix display, from one onward. Let's begin experimenting with the single and dual LED driver

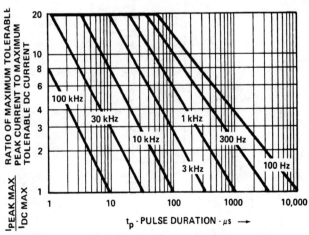

Figure 4-36. *Power strobing and the resulting ratio of DC-to-peak current. (Courtesy of Hewlett Packard)*

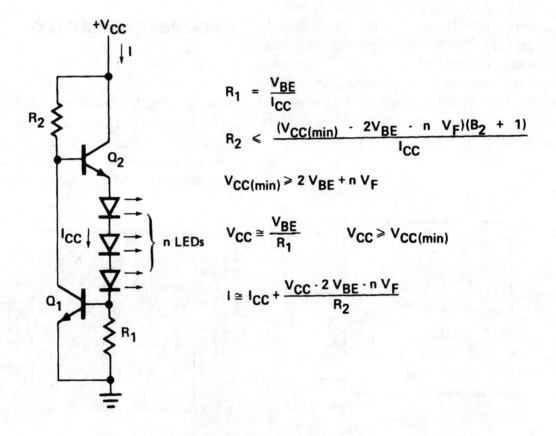

$$R_1 = \frac{V_{CC} \cdot V_{REF}}{1 \text{ mA}} \qquad V_{REF} = 1.5V \text{ for GaAsP LED}$$

$$R_2 = \frac{V_{REF} \cdot V_{BE}}{I_{CC}}$$

$$V_{CC(min)} \geqslant V_{REF} + n \, V_F$$

$$I_{CC} = \frac{V_{REF} \cdot V_{BE}}{R_2}$$

$$I = I_{CC} + \frac{V_{CC} \cdot V_{REF}}{R_1}$$

CAN REPLACE LED$_1$ WITH ZENER DIODE IF DESIRED

$$R_1 = \frac{V_{BE}}{I_{CC}}$$

$$R_2 \leqslant \frac{(V_{CC(min)} \cdot 2V_{BE} \cdot n \, V_F)(B_2 + 1)}{I_{CC}}$$

$$V_{CC(min)} \geqslant 2 \, V_{BE} + n \, V_F$$

$$V_{CC} \cong \frac{V_{BE}}{R_1} \qquad V_{CC} \geqslant V_{CC(min)}$$

$$I \cong I_{CC} + \frac{V_{CC} \cdot 2 \, V_{BE} \cdot n \, V_F}{R_2}$$

Figure 4-37. *Popular transistor driver configurations for multiple LEDs.*
(Courtesy of Hewlett Packard)

Square Wave V_{RMS} = 1.0 Vpeak
Triangular Wave V_{RMS} = 0.816 Vpeak
Sine Wave V_{RMS} = 0.707 Vpeak
Sawtooth Wave V_{RMS} = 0.58 Vpeak

Table 4-2 The Effective Voltage of Wave Shapes Referenced to a Square Wave

Table 4-2. The V_{RMS} values of various wave shapes.

board. (See *Figure 4-23*) Before this though, let's review LED drive formulas, simply using Ohm's Law, for single, multiple and parallel LEDs. (See *Figure 4-35*)

Obviously, your unaided eye can't change how it perceives the color red. We therefore need to enhance its color intensity. Optical glass filtering does this (See *Figure 7-5*), but is expensive. So, let's consider power strobing LEDs. This purposely applies short duration current bursts to the LED, often far in excess of its maximum continuous current. (See *Figure 4-36*) This plots the DC-to-peak maximum current ratio a power strobed LED can endure versus its pulse duration in microseconds, with a 10% duty cycle. Notice how narrower pulses reduce this ratio. Decreasing the duty cycles allow you to apply greater magnitudes of short burst durations of pulsed currents.

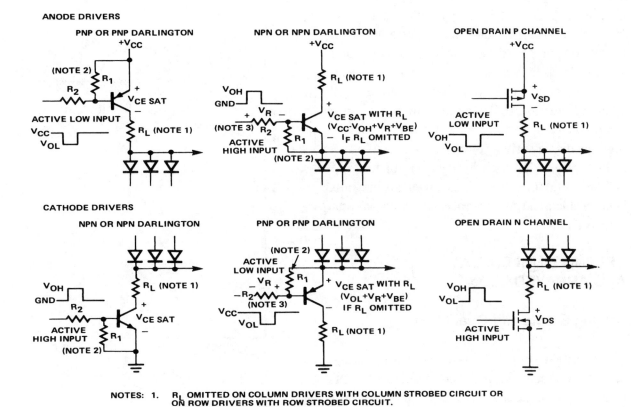

NOTES: 1. R_L OMITTED ON COLUMN DRIVERS WITH COLUMN STROBED CIRCUIT OR ON ROW DRIVERS WITH ROW STROBED CIRCUIT.
2. R_1 CAN BE OMITTED IF TRANSISTOR LEAKAGE IS SMALL.
3. R_2 CAN BE OMITTED IF R_L OMITTED AND TRANSISTOR REMAINS ACTIVE.

Figure 4-38. Driving multiple LEDs with constant current source transistors. (Courtesy of Hewlett Packard)

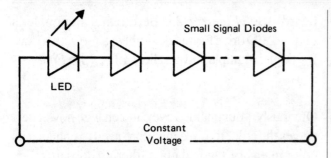

Figure 4-39. *Stabilizing an LED's temperature dependence.*

The three oscillators on the duty cycle controller board (See *Figure 4-31* again) connect to one of the pins of terminal block TB-1 on the Single and dual LED driver board. This power strobes LED DS1 at various duty cycles. Before applying square waves to Q3's base, adjust these circuits' : for 10%, 30% and 60%, and, if you have one, observe their outputs on a scope over five time divisions. Then place a shorting pin in one of the three positions on terminal block TB-1. (See *Figure 4-31*) Pin 1 connects the 10% duty cycle generator, pin 2 connects the 30% duty cycle generator and pin 3 connects the 60% duty cycle generator (oscillator). If you purchase the evaluation board in Chapter 17 from Texas Instruments, you can configure it as a photometer and check your power strobing results against *Figure 4-36*. Power strobing is a common practice in multiplexing (sharing common circuitry) 7-segment display driving circuits. For example, in an eight display design, each display is only on for 12.5% (1/8) the total time. It receives power strobing after being enabled, and does so in a sequential display scanning method.

DRIVING SINGLE, DUAL AND ARRAYED LEDs

Driving Single and Dual LEDs

This board (See *Figure 4-23*) works with inputs from the duty cycle controller and arrayed LED driver board to drive both single and dual LEDs. The four transistors, Q1 through Q4, all purposely have different configurations and polarities for teaching purpose. Specifically, these drive single LEDs, and use PNP and NPN transistors with both emitter and collector drives.

Figure 4-37 provides practical notes on each circuit, plus shows open-drain N and P-channel FET LED drivers. *Figure 4-38* shows constant current LED drive sources which regulate current, regardless of V_CC. Connect the single and dual LED driver board to the BCG board's lines to control these four transistors' bases. Switch their states to observe the LEDs go on and off to check this circuit.

The single and dual LED driver board's remaining circuits drive dual color LEDs. These optoelectronic devices house two different color individual LED substrates within a single two or three-lead package. Our experiments all use two-lead dual LEDs. The usual colors are red and green; however, combinations of these colors, such as orange, do exist, which we later create. Many companies go to great lengths to ensure their dual LEDs have closely matched visible output color intensities. This is not easy since Chapter 1 showed how a green LED appears virtually equal in intensity as a red LED emitting 18 times as much radi-

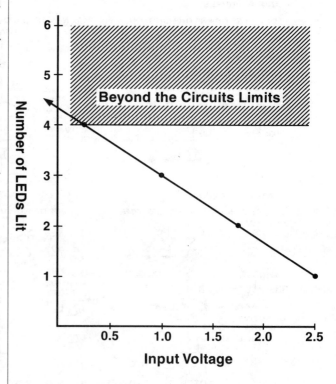

Figure 4-40. *The surprising graph of the number of LEDs lit versus input voltage for the transistor and diode current steering multiple LED "mini" bar graph driver.*

Wavelength, λ in nm	Lumens/Watt
450	26
510	235
555	685
600	430
645	84
700	2.8

Table 4-3. A VLS's color versus its radiated energy.

ant energy. The key factor was the red LED's radiated energy was mostly in the IR range, beyond our vision.

One readily available dual color red/green LED is the Radio Shack catalog no. 276-012. Its data sheet, on this dual LED's thin small cardboard package's reverse side, states it has a 100 mW power dissipation at 30 mA continuous forward current limit with a 2.8 forward voltage. However, it further states the red LED drops 2.0 volts and the green LED drops 2.1 volts, so how do you account for the 2.8 forward voltage drop? There must be an internal series limiting resistor to account for the remaining 0.8 volt (red) and 0.7 volt drop (green) in these LEDs. But its diagram solely shows two LEDs (red and green) in parallel, which defies Kirchoff's Law for total additive series voltage drops! The Lumens intensity is 2.5 mcd with a viewing angle of 54°. Therefore, this is not a very efficient LED. The red LED has a λ = 635 *nm* and the green LED has a λ = 565 *nm*.

Your Eye's Color Preference vs. LED Efficiencies

Calculations within this chapter use 1.5, 1.6 and 1.8 volts as a red LED's forward voltage drop. A GaAsP red LED has a 1.5 volt drop at 10 mA and the extremely old GaP LEDs have a 2.5 volt drop at 10 mA. The multitude of materials and techniques now available to optoelectronic device manufacturers can further increase this range. For example, Dialight's 521 series of discrete LEDs have both red (wavelength λ = 635 *nm*) and green (λ = 565 *nm*). These LEDs reside within a dual color LED, the Dialight 521-9450. The red LED's forward voltage drop is 2.1 volts and the green LED's is 2.3. Their visible light intensities at 10 mA are 2.1 *mcd* (millicandela) and 2.3 *mcd*, respectively. This is exactly 1 *mcd* per each volt of forward drop, but don't draw any conclusions yet. Another LED, the Dialight 521-9432 **high efficiency** red LED, produces 12 *mcd* of visible light at 10 mA with a 2.2 volt drop. This is a far more efficient VLS than even the previous green LED. This is despite your eye being a visible light detector predominantly favoring green, refer to the CIE curve of *Figure 1-2* again. Therefore, carefully read data sheets and don't make rash assumptions about LED colors versus their intensities.

An LED's radiant energy decreases with temperature at approximating -1%/°C. Please realize again that not all radiant energy is visible! The responsivity of your eye to this semiconductor phenomenon, in the red region of λ = 650 *nm*, changes approximately -4.3%/nm. In the green region of λ = 565 *nm*, it changes at a diminished rate of -0.86%/nm. These results are semiconductor and manufacturer driven, but also include your eye's physiology, which determines color sensitivity. Specifically, a red VLS (λ = 625 nm) emits 73 lumens/watt and a green VLS (λ = 565 nm) emits 625 lumens/watt. (See *Table 4-3*)

Placing small signal silicon diodes in series with LEDs stabilize an LED's temperature dependence. (See *Figure 4-39*) These diodes have about a 0.65 volt drop each. Their intrinsic voltage drop is proportional to temperature. An LED's luminance intensity is inversely proportional to temperature, as just stated; therefore, these opposing factors offset or compensate for each other.

When conducting in one direction, one color of the dual LED glows. Driving current in the opposite direction makes the other color LED glow. These two parallel connected (but in opposite polarity direction)

TRUTH TABLE: SIGNAL-LEVEL INDICATOR																							
INPUT V_{in}	A_4										A_5									LED			
	D	C	B	A	$\overline{RBI}$	$\overline{BI}$	a	c	e	g	D	C	B	A	$\overline{RBI}$	a	d	e	f	2	3	4	
below 10%	0	1	1	1	1	1	0	0	1	1	0	1	1	1	0	0	1	1	1			λ	(+6 dB)
10% to 20%	X	X	X	X	X	0	1	1	1	1	0	1	1	0	0	1	0	0	0		λ		(0 dB)
20% to 40%	0	1	1	1	1	1	0	0	1	1	0	1	0	0	0	1	1	1	0	λ			(−6 dB)
40% to 60%	0	1	1	1	1	1	0	0	1	1	0	0	0	0	0	1	1	1	1	ALL OFF			(underrange)
60% to 70%	0	0	1	1	1	1	0	0	1	0	0	0	0	0	0	1	1	1	1	λ			(−6 dB)
70% to 90%	0	0	0	1	1	1	1	0	1	1	0	0	0	0	0	1	1	1	1		λ		(0 dB)
above 90%	0	0	0	0	1	1	0	0	0	1	0	0	0	0	0	1	1	1	1			λ	(+6 dB)

X = don't care   Input voltage V_{in} normalized to full scale at pin 5 of LM3914

Table 4-4. *A truth table of possible color combinations.*

LEDs work on the principle of current taking the path of least resistance. The reverse biased LED's neighbor prevents current flow which allows its parallel forward biased other color LED to pass current. By switching back and forth, at approximately 22 to 30 Hz, or faster, your eye perceives a third color which is a combination of both colors. This rates depends on the LED selected, the drive configuration and its resulting conditions. Gradually varying the dual LED's duty cycle slowly shifts from one to the other LED's color. Chapter 18 presents a color temperature meter project using this same principle and also provides a detailed description of how your eye perceives color.

To demonstrate this, the schematic's transistors, Q3 and Q5, drive a dual LED. Transistor Q3 turns on the red LED and Q5 turns on the green LED. (See *Figure 4-31*) If you observe the reverse of this, you've placed the dual color LED in backwards. Signals BIN 2 and BIN 4, from the BCG board, turn on the base of each transistor. Alternately switching BIN 2 and BIN 4 on and off ensures one LED is exclusively on while the other is off. This is exactly like a flip-flop's Q and ‾Q outputs.

Mixing Dual LEDs' Colors

The next experiment uses J-K flip-flop, U1B, to divide the 150 Hz oscillator's frequency by two. It is composed of two back-to-back two-input NAND gates, U5A and U5B. (See *Figure 4-23*) Toggling only occurs when you hold the J and K inputs and the PR and CLR pins high. Repeat this with the BCG board's BIN 800 and BIN 400 lines. (See *Table 4-4* for a truth table of these four possible combinations.) Square waves have the effect of rapid toggling which causes the dual LED's colors to "mix." It is identical to a painter mixing paints, using just the right proportions of each color. Your results likewise depend on the proportions of the dual LED's red and green intensities. When you hold the J and K inputs low, the BIN 400 signal controls the PR and CLR inputs to the J-K flip-flop to activate its Q and ‾Q outputs. The red **or** the green LED within the dual color LED remains on.

Enhancing a Dual LED's Color Resolution

The next circuit on this single and dual LED driver board operates on the same principle. It only has more sophisticated control to improve color resolution (eight states instead of four). (See *Table 4-4* and *Figure 4-23* again.) You manipulate the control bits of the two 74HCT151 one-of-eight data selectors' address inputs, pins 9, 10 and 11. Their control code selects appropriate data inputs (pins 1 through 4 and pins 12 through 15). These appear at the IC's output (pin 5).

There are two timers, a fast and slow one. The slow 555 timer has an approximate 1 second period or 1 Hz output. The fast timer is the 150 Hz oscillator. You can adjust its frequency by substituting values in the

Frequently it is desirable to operate LEDs on AC power rather than DC. Typically, the power source is 120 VRMS 60 Hz. The most obvious method is to rectify this power with a series diode and use a resistor to limit LED current

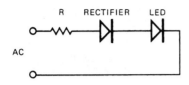

Figure 4-41. *An LED used with half wave rectification.*

0.047 µF to 0.1 µF range for C3. A larger value capacitor produces proportionately slower oscillations. A smaller value capacitor has the opposite effect on the oscillator's frequency. You could build this circuit with either TTL or CMOS double inverted NAND gates. The circuit on the single and dual LED board (See *Figure 4-23* again) centered around the 74HCT00 quasi-CMOS NAND gates in U3 has an obvious time constant (l) determining RC of R23 and C3. However, resistor R22 in such a CMOS oscillator's purpose is to guard against power supply voltage variations which otherwise can cause frequency variations of 30% to 40%. Select this resistor with a value at least twice that of R23 and your variations will remain within the 5% range. The formulas for their periods and frequencies are:

TTL inverter NAND oscillator:
T = 1/f ≅ 1.3 RC

CMOS inverter NAND oscillator:
T = 1/f ≅ 1.4 RC

Try adjusting this oscillator until you reach your own eye's threshold of color detection switching frequency. But, before doing this, you can tell if you need to adjust the frequency of the 1 Hz square wave oscillator by entering the orange code (011) through the three binary lines 40, 80, and 100 (J2-7, J2-8 and J2-9 respectively). This frequency though will be several times above this. Also, try increasing this oscillator's frequency. Its design purposely provides very nearly a 50% duty cycle (symmetrical square wave); other-

wise, you would not achieve a realistic orange color. Duty cycle is crucial since we conduct in alternate directions. Step through the eight binary combinations to see for yourself how this simple circuit yields impressive visual effects.

A 50% duty cycle is crucial; however, using a variable duty cycle square wave purposely produces unequal conduction of current in one direction versus the other. This yields a solid red, slowly but continuously changing in hue until it reaches a solid green (using the 1 Hz slow oscillator.) It then incrementally reverses this solid green back to red. (See *Figure 4-31*) You are highly encouraged to also breadboard this circuit.

Proving the Eye's Bias Toward Green

To demonstrate the eye's receptiveness to different colors in an experiment, refer to the three different color LEDs, DS2 through DS4 in *Figure 4-31*. Adjust their pots until all three appear equally bright to you! This may never occur though since green always tends to appear brightest. Regardless, after your "best effort" adjustments, turn the board's power off and measure the series resistance of all three color LEDs. Record and analyze your results by referring to their data sheets' brightness in Lumens. See *Table 4-3* again for a guide.

Driving Arrayed Series Connected LEDs

The five transistors, Q7 to Q11, along with their current steering and blocking diode inputs (See *Figure 4-31*) comprise an unusual bar graph driver. This circuit works just the opposite of normal LED array drivers which light LEDs from the bottom up with increased input voltage. Increasing this circuit's input voltage though activates **fewer** LEDs. Conversely, decreasing input voltage **increases** the number of lit LEDs. *Figure 4-40* graphs this. Resistors R53 through R57 supply sufficient base current to Q11 through Q7 to drive them into saturation. When the input voltage decreases, through pot R50, the current in R57 divides, diverting through D5 to ground if pot R50 is at its midway position. This turns off Q11 and illuminates LED DS17. You may use the previous equation, de-

Here a capacitor is used to control LED current and a shunt silicon diode provides rectification.

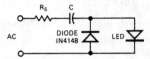

Figure 4-42. *An LED using a dampening capacitor for AC's effects.*

scribing a straight line, to derive the following equation with "DL" representing the number of LEDs lit in this circuit. This is a good exercise for you:

$$Y = mx + b$$

$$V_{IN} = 3.25 - 0.75 \text{ (DL)}$$

AC LED Applications

Don't overlook these just because an LED draws a maximum of just 20 mA and drops approximately 2.0 volts, depending on its type. *Figure 4-41*'s LED uses half wave rectification and a series limiting resistor. *Figure 4-42* improves upon this "brute force" method with a series capacitor and a parallel diode for rectification. *Figure 4-43* plots this capacitor's value versus the LED's current at 120 V_{RMS} at 60 Hz. This RC combination dampens turn on voltage surges and "in

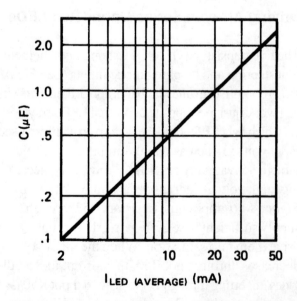

Figure 4-43. *The capacitor values required for various LED current flows.*

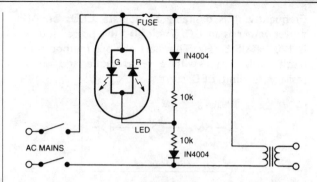

Figure 4-44. *A dual color LED used in an AC application, which indicates the status of both the AC input fuse and the AC lines.*

rush" current. (See "dv/dt" and "snubber" circuits both in Chapter 14 and in the burglar baffler project in Chapter 18).

Figure 4-44 shows the last LED AC application, but this time using a dual color LED to monitor the AC power to a circuit and indicate if the fuse has blown. Regardless of the fuse's status, the GREEN LED is

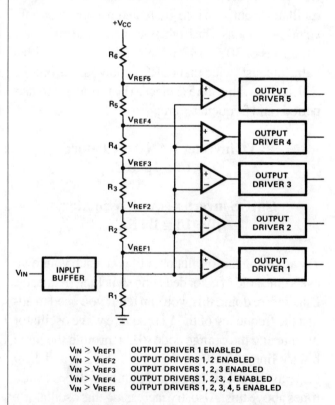

$V_{IN} > V_{REF1}$	OUTPUT DRIVER 1 ENABLED
$V_{IN} > V_{REF2}$	OUTPUT DRIVERS 1, 2 ENABLED
$V_{IN} > V_{REF3}$	OUTPUT DRIVERS 1, 2, 3 ENABLED
$V_{IN} > V_{REF4}$	OUTPUT DRIVERS 1, 2, 3, 4 ENABLED
$V_{IN} > V_{REF5}$	OUTPUT DRIVERS 1, 2, 3, 4, 5 ENABLED

Figure 4-45. *An example of op amp comparators without a committed output. (Courtesy of Hewlett Packard)*

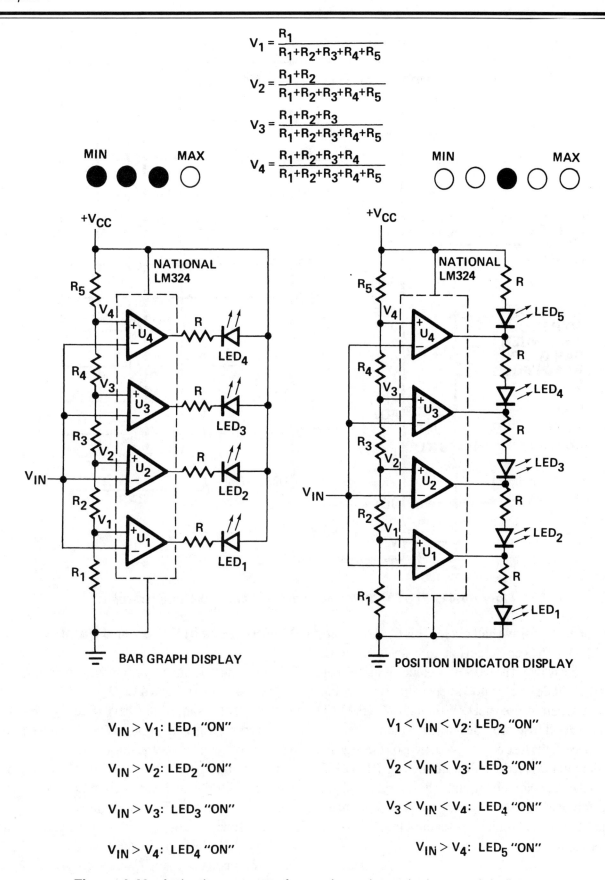

$$V_1 = \frac{R_1}{R_1+R_2+R_3+R_4+R_5}$$

$$V_2 = \frac{R_1+R_2}{R_1+R_2+R_3+R_4+R_5}$$

$$V_3 = \frac{R_1+R_2+R_3}{R_1+R_2+R_3+R_4+R_5}$$

$$V_4 = \frac{R_1+R_2+R_3+R_4}{R_1+R_2+R_3+R_4+R_5}$$

BAR GRAPH DISPLAY

POSITION INDICATOR DISPLAY

$V_{IN} > V_1$: LED$_1$ "ON"

$V_{IN} > V_2$: LED$_2$ "ON"

$V_{IN} > V_3$: LED$_3$ "ON"

$V_{IN} > V_4$: LED$_4$ "ON"

$V_1 < V_{IN} < V_2$: LED$_2$ "ON"

$V_2 < V_{IN} < V_3$: LED$_3$ "ON"

$V_3 < V_{IN} < V_4$: LED$_4$ "ON"

$V_{IN} > V_4$: LED$_5$ "ON"

Figure 4-46. *Activating outputs of a moving point <u>and</u> a bar graph indicator. (Courtesy of Hewlett Packard)*

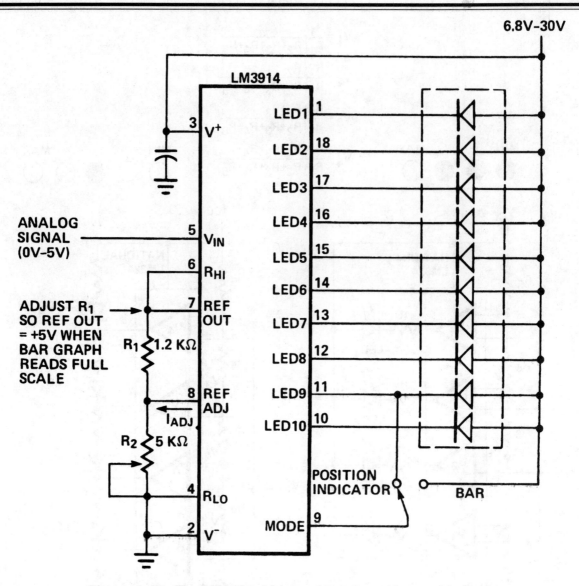

Figure 4-47. *The LM3914 driving a 10-segment bar graph display.*

ON for each positive half cycle. This 60 Hz exceeds an LED's flicker detection rate so it **appears** continuously ON. The basis of operation is the average rectifier diode (the 1N4001 here), draws more than 10 mA so current does not flow through the RED LED since a good fuse provides it with a path of far less resistance. If the fuse does blow, current on the negative portions of the AC cycle flow through the RED LED. This causes a color mixing and yellow results. The AC transformer draws some current; therefore, you should adjust the current limiting resistor of the RED LED, accordingly.

Driving LEDs in Series and Parallel

The following two "mini" bar graphs use quad op amps, U2 and U4, to drive 5 and 4 LEDs respectively. Each uses exactly the same input drive (See *Figure 4-45*), but different LED activation methods. Both circuits are improvements over the unorthodox top-to-bottom illuminating transistor and diode blocking and steering LED driver in *Figure 4-31*, based on the equation of a straight line. These designs use fewer components. Each IC contains four op amps and **they** drive the LEDs, instead of transistors. The first circuit, built around quad op amp U4, is a *bar graph* indicator. The second circuit, built around quad op amp U2, is a *moving dot* or moving point indicator.

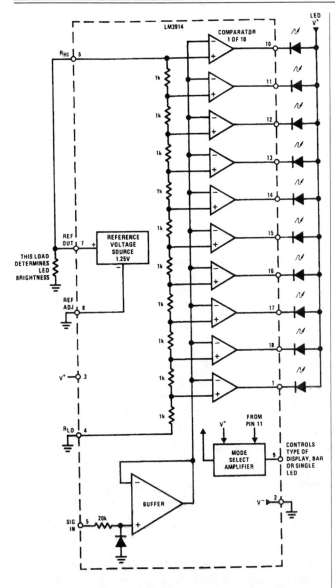

Figure 4-48. *A simplified block diagram of the LM3914 IC. (Courtesy of National Semiconductor)*

The Bar Graph Mode

Figure 4-46 illustrates input voltage magnitudes and their subsequent activated op amps. Note the activated comparators' lit LED(s) react differently. A black filled in dot in *Figure 4-46* represents a lit LED. The circuit on the left is the bar graph indicator. This display indicator's voltage dividing resistors, R45 to R49, equally divide the 12 volts into five 2.4 volt successive increments which appear on each of U4's four noninverting (+) inputs. Pot R40's setting determines

how much voltage appears on all four of U4's inverting (-) inputs.

Any op amp, in a comparator configuration like these, compares the magnitude of voltage on its two inputs. If the result is them inevitably being unequal, it drives the op amp output to its "rails." This means its output either goes to its lower voltage supply (ground in our case) or to its upper supply voltage (+12 VDC in this case). If the op amp's inverting (-) input has a greater voltage than its noninverting (+) input, its output goes to a virtual ground level (very nearly 0 volts). Conversely, if the op amp's noninverting (+) input has a greater input voltage on it than the inverting (-) input, its output goes to a virtual V_{DD}, or in this case, very nearly +12 volts. As the inverting (-) inputs on each U4 op amp surpass each successive 2.4 volt increment threshold, it drives that op amp, within the quad op amp IC, low. This sinks (draws in) current through its output LED. This current is: [12 volts - the LED drop (1.8 volts for a red LED)] divided by 1 kW (the values of current limiting resistors R41 to R44). This calculation yields a 10.2 mA current for each LED. This is quite adequate to drive red LEDs.

The first LED, (DS10), turns on as U4A's noninverting (-) input (pin 11) exceeds 2.4 volts (V_1 in *Figure 4-46*). LEDs DS10 and DS11 turn on as U4's noninverting input (pin 8) exceeds 4.8 volts (V_2 in *Figure 4-46*). LEDs DS10, DS11 and DS12 all turn on as U4's noninverting input (pin 6) exceeds 7.2 volts (V_3 in *Figure 4-46*). All four LEDs turn on as U4's noninverting input (pin 4) exceeds 9.6 volts, (V_4 in *Figure 4-46*). Again, note the black or "filled" in dots representing lit LEDs.

The Moving Dot Mode

This indicator's display uses the other quad op amp, U2. Using an LM324 quad single supply voltage op amp for U2 is one of many acceptable op amps. This design, up to the op amp outputs, is again **exactly** like the U4 based circuit. A five resistor voltage divider, composed of resistors R29 to R33 again equally divides the 12 volts into five 2.4 volt successive increments. Again, pot R28's setting determines how

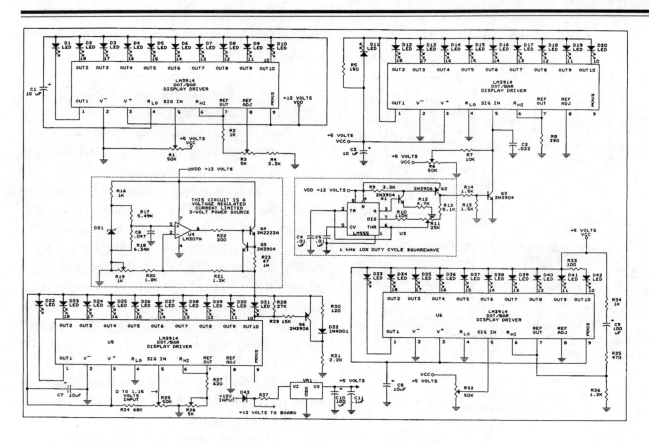

Figure 4-49. *The bar graph driver board schematic.*

much input voltage appears on the four inverting inputs of U2. As the input voltage surpasses each incremental 2.4 volt successive threshold, it drives its op amp's output low. This again sinks, draws in, or provides a path to virtual ground for current flowing through its LED. Each of the five LED's series current (10.2 mA) limiting resistors (R23 through R27) are again 1 kΩ. This though is where these quad op amp indicators' driver circuits **differ**.

The U2 based circuit has all its LEDs in series with one another. This time, as you reach each 2.4 volt threshold, it activates that op amp's LED. However, all LEDs below any activated op amp's LED turn **off**. The first op amp, U2D and its LED, DS5, naturally though does not turn on or off any LEDs below it, since none exist. This concept of no LEDs turning on below the highest activated op amp(s) is the basis of a moving dot indicator's operation! This occurs because it prevents any current flow through the LEDs and their series limiting resistors below this one and only activated op amp and its driven LED. Once a U2 op amp

output goes low, the series current flow ceases, due to a lack of driving potential (0 volts) for the LED directly below it. This is why it is called a single moving dot, and not a bar graph driving display. The first quad op amp, U4, has all its output LEDs in **parallel** with +12 volts and is responsible for this unique characteristic. The second quad op amp, U2, has all its output LEDs in **series** with +12 volts. Scrutinize the outputs, in the form of filled in dots in *Figure 4-46* one last time.

Subtle Design Considerations

These two circuits assumed we used all red LEDs; however, you may place different color LEDs, in any number or combination, in either U2 or U4's LED output array. Realize, however, that you will probably have to change the current limiting resistor's value for each new color. Consult the LEDs' data sheets for current versus visible output intensity, expressed in Lumens. Then select series current limiting resistors to match the same quantity of Lumens for each color

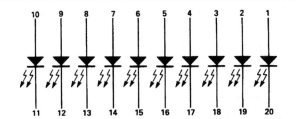

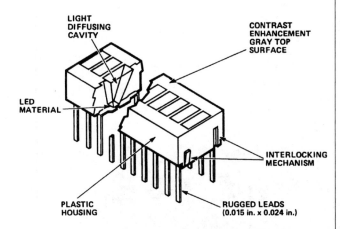

Figure 4-50. The internal structure and schematic of a typical 10-segment bar graph display.

LED. Refer back to your experiment with three different colors of LEDs.

Your voltage divider resistors' tolerance is not as crucial as their matching ratios. These are what divide voltages into equal increments. They should be as close as possible to 1:1 throughout. Manufacturers of SIP and DIP resistor packs have resistor arrays with critically matched resistor ratios, rather than just primarily stressing absolute tolerance accuracies of resistor values alone. Another important ratio is the matching of resistor temperature coefficients so drifting all occurs in the same direction and magnitude, which wouldn't adversely affect a voltage divider.

Preparing for the LM3914

Emphasizing these two ratios is vital for precision voltage divider applications like these. The LM3914 overcomes these three factors:

1. Series or parallel op amp comparator output resistor configurations.
2. Voltage divider resistance ratios.
3. Voltage divider resistance temperature coefficients.

The LM3914 demonstrates how much thought went into its design by providing all three of these "subtle" details. Specifically, its built-in voltage reference and 10-step voltage divider relieve our previous concerns over both resistance temperature coefficients and resistance value matching ratios, and, it has internal current limiting output resistors. All exist on the same substrate; therefore, its tight tolerance laser trimmed internal voltage divider resistors, all therefore inherently have closely matching resistance ratios and temperature coefficients.

Our previous quad op amp examples of driving arrayed LEDs in clusters showed you how bar graph and moving dot LED indicators work, and how they differ! Admittedly, modern designs would never use as many discrete components as our quad op amp series and parallel LED drivers did. The previous paragraph showed how IC manufacturers place as many passive components as possible on monolithic substrate to minimize parts count, temperature drift and costs. This is equivalent to 2 1/2 quad op amp ICs, but the LM3914 also relieves you of placing series or parallel resistors on the output. One resistor regulates current drive to all the LM3914's output LEDs. Wow, what a relief from our quad op amps' current limiting resistors in parallel and series! This not only eliminates 10 LED series limiting resistors but also allows operation from 3 volts or less, typical of modern laptop computer's 3.3 volt logic.

BAR GRAPH DRIVING

The LM3914 Bar Graph Display Driver

The LM3914 senses input voltage levels to 35 volts, above and below ground, referenced to its precision divider to linearly drive ten output LEDs. Nonlinearity is typically better than 0.5%, even over a wide temperature range. Within this three IC series, the LM3915

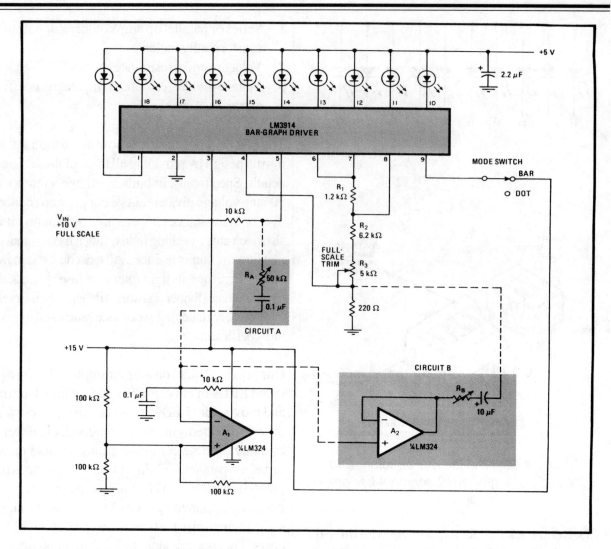

Figure 4-51. *Expanding a bar graph driver's resolution to the limits of your eye's ability to differentiate relative brightness levels.*
(Courtesy of EDN/Chaners Publications)

is the LM3914's equivalent, with a logarithmic output. The LM3916 has an identical output to the LM3915; except, its logarithmic output is referenced to 600 ohms. This corresponds to volume units (vu) for audio applications.

A single MODE pin (9) (See *Figure 4-47*) changes this IC's display from a moving dot (low) to a bar graph (high). This lights all LEDs, up to and including the LED indicating the input voltage threshold. In the dot mode, only the last activated LED in the "stack" lights to indicate its input voltage. Like our previous quad op amp moving dot circuit, all segment bars or LEDs below this activated output remain off.

The LM3914 provides visual alarms, dimness control and the ability to cascade multiple driver ICs, forming up to a 10 segment bar graph. (See *Figure 4-48*) Resistor R1 is the resistance between pins 7 and 8 and R2 is the resistance from pin 8 to ground. The resistance from pin 7 to ground determines brightness which is inversely proportional to this :.

The LM3914 More Efficiently Driving Bar Displays

The first LM3914 application, U1 in *Figure 4-49*, is the basic configuration and simplest drive technique for 10-LED arrays. The 20-pin bar graph display it

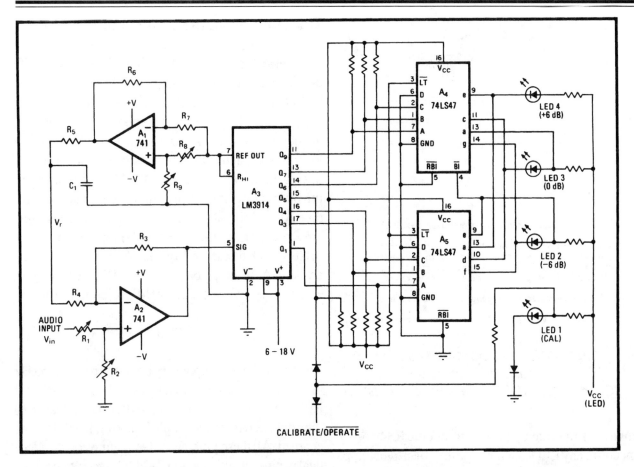

Figure 4-52. An LM3914 and 7447 based "compacted" log output circuit, with its output displaying dB (decibels 6 dB apart). (Courtesy of EDN/Chaners Publications)

typically drives (See *Figure 4-50*) has 10 uncommitted LED anodes and cathodes. Pot R1 controls the applied input voltage and pot R3 is the REF ADJ (Reference Adjust) pin. It conforms to the following equation with R1 in this equation representing U1's R2 plus R3, up to its wiper. R2 in the equation represents the resistance from R4 to U1's R3 resistance, up to its wiper, from the direction:

$$V_{OUT} = V_{REF} \ (1 + R1/R2) + I_{ADJ} \ R2$$

As its unique input, U3 in *Figure 4-49* uses a 555 timer's approximately 10% duty cycle output to illuminate the LEDs in an "exclamation point" attention getting fashion. You permanently hard wire **ON** the lowest LED, D11. The other nine LEDs above D11 ramp up, following the amplitude of the input voltage, and flash. To your eye, this appears to increase reso-

lution while providing an overvoltage indication. The 555 timer's output modulates the input signal and pot R11 controls its duty cycle.

The next LM3914 application, U5 in *Figure 4-49*, is an indicator and alarm which changes from the dot to bar mode after reaching full scale. Detach the input to the dot-bar switch from the anode of the tenth LED, D31, and attach it to another LED's cathode. Do this with a shorting jumper on one of the terminals of its 10 terminal strip. The display again changes modes as your selected new LED reaches its input voltage triggering threshold. The circuit uses a 3 volt power source designed around op amp U4.

The last LM3914 application, U6 in *Figure 4-49*, is a bar display and alarm, again attaining full scale and flashing as a bar graph display; however, you may again

Percentage ON (Duty Cycle)	Resultant Color of the Dual LED
80%[1]*	Darker Green
70%	Yellowish-Green
60%[1]	Darker Yellow
50%	Orangish-Yellow
40%[1]	Orangish-Yellow
30%	Redish-Orange
20%[1]	Red
10%	Red

[1] These represent the four-step color resolution

Note: If these color appear in a reversed order you have the 2-lead dual color LED in backwards, however; it hardly matters since it still illustrates the principle of color resolution

* This recycles after completing its cycle

Table 4-5 Four and 8-step resolution color versus duty cycle

Table 4-5. *The LM3914 dB output circuit's truth table. (Courtesy of EDN Chaners Publications)*

move the full scale LED and jumper the R34-to-C9 junction to any LED's cathode via the 10-terminal strip. Adjust pot R32 until you attain the input voltage threshold. Its unique feature is voltages, up to and beyond this jumpered LED's cathode, flash the display.

"Dithering" the LM3914

You can realize an expanded scale single LED bar graph display with a scanning type "dithering" or modulated circuit, developed by Robert Pease of National Semiconductor. It approaches infinite resolution! It uses a triangular wave shape oscillator and either one of the two simple dotted line RC networks, designated "A" or "B." (See *Figure 4-51*) These capacitively couple, attenuate and superimpose their output on the LM3914's input, pin 5. Set the 50 kΩ pot so each desired incremental input change causes successive bar graph LED elements to glow, gradually spreading across the whole display. If your input voltage creates heavy loading, or isn't linear, these AC coupled superimposed input signals yield erroneous results. Use circuit "B" in this case to buffer the triangular wave oscillator.

Transition from the LM3914 to the 7447

Figure 4-52 combines both a 74LS47 7-segment LED display decoder/driver IC (discussed next) and the LM3914 linear output LED bar graph display driver. This circuit's though output displays AF (Audio Frequency) asymmetrical signals in dB (decibels), or in a logarithmic manner. If you can tolerate diminished output resolution, since not all LM3914 outputs are used, the output deviates from center scale to accommodate AF AC signal swings. Op amp A1 applies a reference voltage to the inverting (-) input of A2. This offsets both itself and the LM3914 by a desired amount. Assuming you make R6 = R7, the following equation describes the LM3914's reference voltage:

$$V_r = 1.25[-2R9/(R8 + R9) + 1]$$

You can make the output vary linearly, from -1.25 kHz to +1.25 kHz, at any value, by adjusting pots R8 and R9. The two 74LS47 ICs have their lamp test (LT) pins disabled and their ripple blanking In (RBI) pins enabled. All these pins, and other features on the identical in pin-out and function 7474 IC, are covered next. The A5 74LS47 has its A to C inputs ascend in the correct direction, but with a skewed order, since they accept the LM3914's Q1, Q3 and then Q4 outputs. The A4 74LS47 has its A to C inputs descend in what appears to be the wrong direction from LM3914's Q9, Q7 and then Q6 outputs. *Table 4-5* is a truth table for this circuit. LED1 lights to indicate calibration from the CALIBRATE/OPERATE line. Be careful the noise from LED1 does not reach a logic low threshold for the LS type TTL (transistor-to-transistor logic) 74LS47's blanking input (BI), pin 4. The LS TTL logic is low-powered Schottky and draws less current than TTL, but is no where nearly as efficient as CMOS logic.

SEVEN-SEGMENT LED DISPLAYS

The Basis of Operation

Seven-segment LED displays use LEDs for each segment and/or decimal point(s). You can stack these rectangular packages side-by-side. In most cases, not all

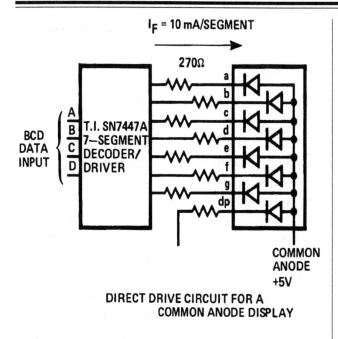

I_F = 10 mA/SEGMENT

270Ω

BCD DATA INPUT

A
B
C
D

T.I. SN7447A 7-SEGMENT DECODER/ DRIVER

a
b
c
d
e
f
g
dp

COMMON ANODE +5V

DIRECT DRIVE CIRCUIT FOR A COMMON ANODE DISPLAY

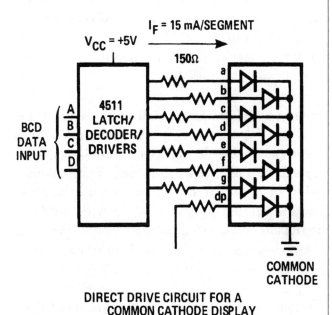

I_F = 15 mA/SEGMENT

V_CC = +5V

150Ω

BCD DATA INPUT

A
B
C
D

4511 LATCH/ DECODER/ DRIVERS

a
b
c
d
e
f
g
dp

COMMON CATHODE

DIRECT DRIVE CIRCUIT FOR A COMMON CATHODE DISPLAY

Figure 4-53. *Common cathode and anode 7-segment display drives and their internal connections.*

display pins are brought out. Since 7-segment LED displays have seven or more LEDs electrically connected at either their cathodes or anodes (See *Figure 4-53*), there must be at least one external pin for this common connection. This "wiring" redundancy provides alternate current sharing paths to protect the

display's wire bond-to-substrate integrity. Over 130 mA of current can flow with all segments and decimal point(s) lit! Sometimes there is a right or left hand decimal point "eighth" segment. (See *Figure 4-54*) Other specialty displays have less than seven segments, such as a +/- 1 display, with one or two decimal point(s) to the left, right or on both sides. (See *Figure 4-55*) Decoder/driver ICs drive LED displays by selectively lighting their segments. If you ground a particular LED segment in a common anode display, it lights. If you ground all seven segments in a common anode display, through their cathodes, all segments light. This forms an "8" pattern. Small letters "a" through "g," denote these seven segments. (See *Figure 4-56*) Conversely, in a common cathode configuration, applying a voltage to an LED's anode, (with the cathode grounded) lights that display's segment.

You must use a common anode driver with a common anode display or a common cathode driver with a common cathode display. Your experiments use common anode displays which are the more prevalent style connection. If you want to mix driver and display types, you'll need an inverter at each output of the IC. This is an otherwise unnecessary IC though. Don't mix IC driver logic families. Compensating for this mismatch requires far too many components.

Figure 4-54. *Seven-segment LED displays with decimal points.*

Figure 4-55. *Specialty non-7-segment LED displays with decimal points.*

One of the oldest display drivers is the TTL 7447, but its features, short of internal current limiting resistors, are representative of most modern drivers. After getting acquainted with the 7447, you can transfer these basic concepts and knowledge to the myriad of decoder/drivers on the market today. The 7447 is an open collector IC accommodating up to +15 VDC on its outputs and has leading/trailing zero suppression, lamp test and intensity modulation capabilities. Lines A through F have bubbles indicating active low outputs. (See *Figure 4-27* again) The 7448 and 7449 are the identical in function common cathode versions of the 7447. A common CMOS logic decoder/driver is the CD4026. This modern IC has built in series current limiting resistors. Motorola used to manufacture a really handy BCD-to-hex 7-segment decoder/driver, the MC14495.

When active (low), current flows through the display, the series limiting resistors, and to its output. When active, these outputs are only a few tenths of a volt above ground. The 7447's input side has four inputs, 1, 2, 4, and 8. These are binary inputs for decimal numbers, up to and including 9. If you apply numbers greater than 9, strange unrecognizable segments occur, which you'll soon learn how to correct.

Your first decoder/driver experiment examines the control lines: LT for lamp test, RBI for "ripple blanking in" and BI/RBO for "blanking in/ripple blanking out". If you want to enable output functions 0 through 15, you must open or tie the blanking input high (BI pin 4 on IC U1 and BIN 10 from the BCG board). One of three different signals on TB#1 drives the ripple blanking input (RBI) pin 5 on IC U2. Leave pin 5 open or high if you don't desire decimal zero blanking.

The first position on TB#1 (pin 1 to pin 2), accepts logic inputs from the BCG board. The second position (pin 3 to pin 4) is from the 555 timer, U1. The third position (pin 5 to pin 6) accepts one-shot pulses triggered by the free running or astable oscillator composed of NAND gates U3. (See *Figure 4-27*) Position #4 completes the circuit, connecting D1 between pins 11 and 13 of the 7447.

A low on the blanking input (BI pin 4), shuts all display segments off, regardless of their input states. When the ripple blanking input (RBI pin 5) and inputs A, B, C, and D are low, and the lamp test (LT pin 3) is high, all segment outputs turn off and the ripple blanking output (RBO pin 4) goes low. An open or high blanking input/ripple blanking output (BI/RBO), pin 4, and a low lamp test (LT) input, pin 3, turn on all display segments.

Tie pins 3 and 4 high, move the shorting pin to #2 on TB#1 and adjust pot R43 up and down. Observe the changing display brightness. You are pulse width modulating (PWM), a phrase or varying the 555 timer's duty cycle from approximately 10% to 90%. The greater the duty cycle, the brighter the display. Move the shorting pin to position #3. Substitute an automatic brightness adjusting photocell, R38, for the manual brightness adjusting one-shot's pot. *Figure 4-57* is another circuit which also uses a photocell's light sensitive resistance, but this time to change the pulse width of a retriggerable one-shot IC. This circuit does a "pulse stretching". A one-shot normally produces just

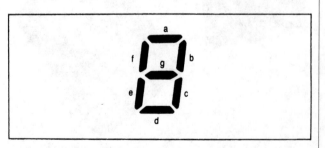

Figure 4-56. *A 7-segment LED display's segment designators "a" through "g."*

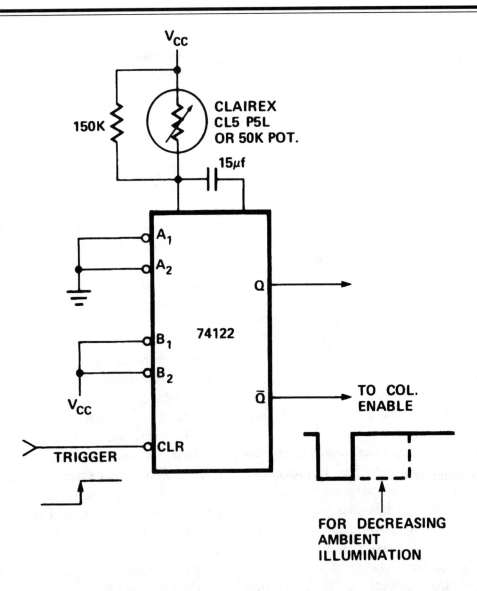

Figure 4-57. *A photocell's light sensitive resistance controlling a one-shot IC's pulse width*

one pulse; however, retriggerable versions feed their outputs back around to serve as "triggers" for their inputs. A typical CMOS retriggerable one-shot is the CD4528 (a CD4098 is an identical IC in function). The photocell's TO-5 case has a clear optical glass window. Cup your hands around this photocell or totally block its light and notice the duty cycle decrease to its minimum. This is expected since dark ambient lighting does not require as brightly illuminated VLSs. Shine a flashlight or strong VLS into the photocell. If you have one, observe a maximum duty cycle output at pin 6 of IC U7 with a scope over five time divisions. As expected, display DS1 brightens.

Limitations of the 7447 and its Solutions

One annoying trait is incompletely forming numerals "6" and "9". The upper and lower bars respectively do not light. Place a diode, D1, between the 7447 pins 13 and 11 to remedy this by moving the shorting pin to position 4 of terminal block TB#1. Apply a binary "6" (110) with the BCG to lines BIN 4, 2 and 1 respectively. Remove the shorting pin and observe the upper segment bar in the "6" go away. Another cure to the missing segment bars in both the "6" and "9" is inserting two open-collector NAND gates in the in-

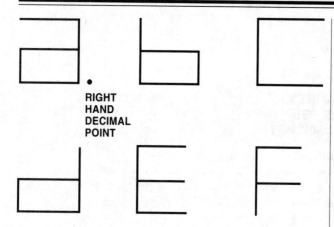

RIGHT
HAND
DECIMAL
POINT

Table 4-6. *Hex numerals formed beyond decimal 9.*

put. (Refer to IC U6 in *Figure 4-27*.) Apply a binary "6" and "9" with the BCG board and observe this "fix."

A second annoying 7447 feature is its inability to decode and display hexadecimal equivalents of decimal 10 through 15 (A to F in hex). The circuit in IC U4 partially solves this by representing lower case numerals 10 through 13 or A through D in hex. This requires a display with a right hand decimal point. The decimal point forms the small lower right hand "tail" on a lower case "a." (See *Table 4-6*) To form a lower case "c", light the display's "d," "e," and "g" segments by running current through the 51 ohm resistor, R13. The A and B inputs control segments "b," "f," "c," and the

right decimal point. This last circuit is a common anode display design; however, you could use a common cathode display by tying the 51 ohm resistor to the +5 volt supply. This suppresses blanking on illegal input combinations. Try these four possible input combinations.

Detecting Decimal Numbers From Segments

You may some time need to convert a number in 7-segment format to decimal with just four or less input gates. This might coincide with lighting a warning VLS on a control panel corresponding to a displayed number. For example, *Figure 4-27* shows three ways to detect "0".

Segments from a scanned source, such as the earliest calculator ICs, required an additional input and only detected the desired position. The next generation calculator ICs blanked their displays during computation. Just in case you want to experiment with an old calculator in your junk box, realize calculator ICs drive LED segments without external current-limiting resistors. They use current sources (FETs) for limiting current. These older calculators' LED displays were quite small and shaped with adjacent dome shaped lens, most often covered by a colored plastic light filter. Their voltage output drop is just across the LED (typically 1.8 volts for a red LED).

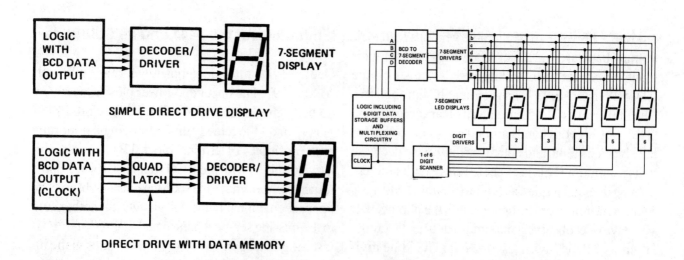

Figure 4-58. *A circuit to multiplex LED 7-segment displays.*

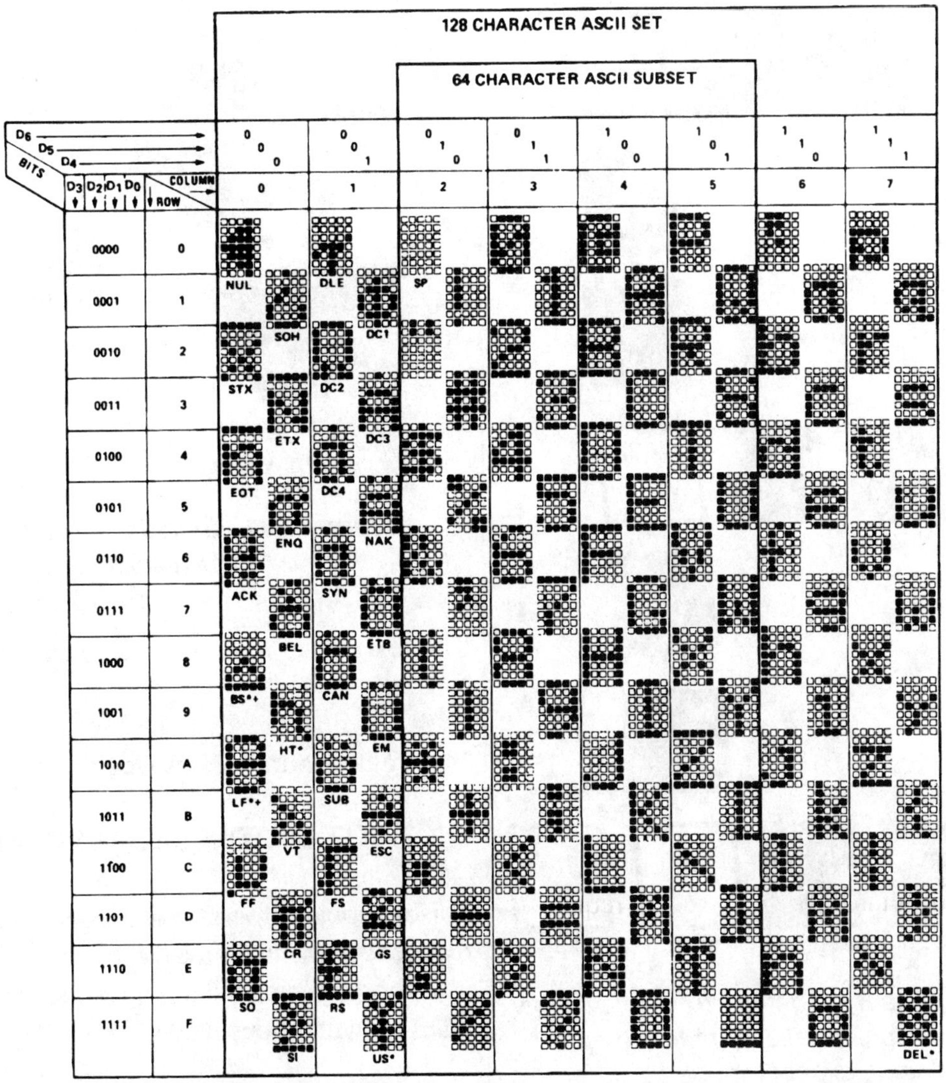

*DISPLAY COMMANDS WHEN USED IN LEFT ENTRY
+DISPLAY COMMANDS WHEN USED IN RIGHT ENTRY

Figure 4-59*. The 5 x 7 dot matrix's 128 possible ASCII characters and symbols.*

IC U6 and the three gates of IC U5 detect the numerals "1" and "5", as well as the minus "-" sign. Apply a "1" and watch LED D2 illuminate. Apply a "5" and watch LED D3 illuminate and apply any number with the segment "g" active and LED D1 illuminates.

Multiplexing 7-Segment Displays

This practice uses power strobing or, applying high currents at low duty cycles, as we've done to single LEDs. This takes advantage of an LED's higher optical efficiency at higher currents and lower power dissipation than operating with DC levels. Your eye's tendency to spot saturate also helps! *Figure 4-58* is a block diagram showing a six 7-segment display with storage buffer memories, a 1 of 6 scanner to scan across either the common cathodes or common anodes of the displays, depending on their type and a normal BCD-to-7-segment decoder with segment drivers or current buffers on each of its output.

On the first clock pulse the multiplexing logic selects storage buffer number 1. The digit scanner only enables display digit number 1 and allows this digit's appropriate segments to light. This same incident occurs again on the next digit, number 2, and sequences in order to digit number 6. It repeats in a display number 1 to 6 cycle. To optimize the power strobing rate,

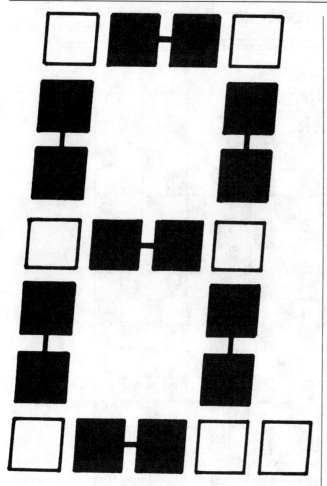

Figure 4-60. *The 5 x 7 dot matrix display's row and column configurations (note the right column is not used, except for one LED, for this custom pattern)*

you should scan at a refresh rate of 1 kHz or above. Important parameters in this type drive circuit include:

1. The refresh rate, f.
2. The quantity of displays, N.
3. The duration of the ON-pulse, t_p.
4. The blanking time, t_b, between digits.

The refresh rate is the reciprocal of the refresh period, as the following equation describes.

$$\textbf{Refresh Period} = \textbf{p} = \textbf{1/f}$$

Figure 4-61. *Single 5 x 7 dot matrix LED displays.*

The next equation describes the maximum ON-pulse time as:

$$\textbf{Maximum ON-pulse time} = \textbf{t}_\textbf{p} = \textbf{p/N}$$

5 X 7 DOT MATRIX LED DISPLAYS

This significant progression in VLS drive complexity requires commercially available memories. Therefore, your thorough understanding of these type displays relies upon first, your understanding of how memories

Figure 4-62. *A small 4-digit 5 x 7 dot matrix LED display.*
(Courtesy of Hewlett Packard)

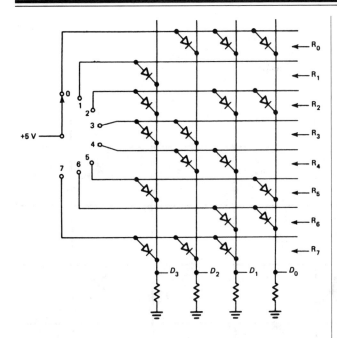

Figure 4-63. *The simplest possible memory, a diode ROM.*

Register	Word	Address
R_0	0111	0
R_1	1000	1
R_2	1011	2
R_3	1100	3
R_4	0110	4
R_5	1001	5
R_6	0011	6
R_7	1110	7

Table 4-7. *The diode ROM memory's contents and its addresses.*

function (*Figures 4-59* to *4-62*), which the following memory tutorial provides. It begins with the simplest memory imaginable.

Basic Memory Concepts

A certain type of MOS transistor universally comprises a semiconductor memory; however, we'll begin with an elementary 8-bit wide diode ROM memory, composed of diodes and a multi-position mechanical switch. Its principal disadvantage is not having two conducting states, which a transistor provides. This obviously impractical memory serves us well though for demonstrating the key memory concepts of:

1. Registers.
2. Addressing.
3. Decoding.
4. Programming.

Registers

Each horizontal row of the diode ROM in *Figure 4-63* serves as a register. A register is an orderly grouping of memory elements or "cells" working together. The simplest register does nothing more than store a binary word or group of bits (usually eight in quantity,

called a byte). Other special registers, not used here, modify their contents by e.g. shifting (rotating data in either direction) or adding a one to it. *Table 4-7* lists the diode ROM memory's data by address.

Addresses

An address in memory closely corresponds to a physical street address. It allows an orderly search and assigns a numerical value, (unique address), to each memory cell. Most commercially available memories are 8 or 16 bits wide. A memory's capacity is specified in kilobytes or 1,024 bytes. The term kilo normally means 1,000; however, since memories are based on binary numbers, $2^{10} = 1,024$ and that is close to, but not exactly 1,000 bytes. Our ROM diode circuit has addresses 0 to 7, each corresponding to a position on the rotary mechanical switch. Actual IC memories have built in decoders.

Decoding

Our ROM diode memory is impractical for a second reason, it would require too many input pins, one for each address. *Figure 4-64* is an IC ROM with on-board decoding (address selection) and an identical memory content of the diode ROM. Input pins A_0, A_1 and A_2 supply the stored word's binary address. For example, with address 101, the decoder applies a high to register R_5 and the contents are 1001. Most small IC memories have eight or more input lines, allowing decoding of up to 256 words and registers (2^8). By using decoding we eliminate the unrealistic and cumbersome need for 256 input lines, as our diode ROM requires.

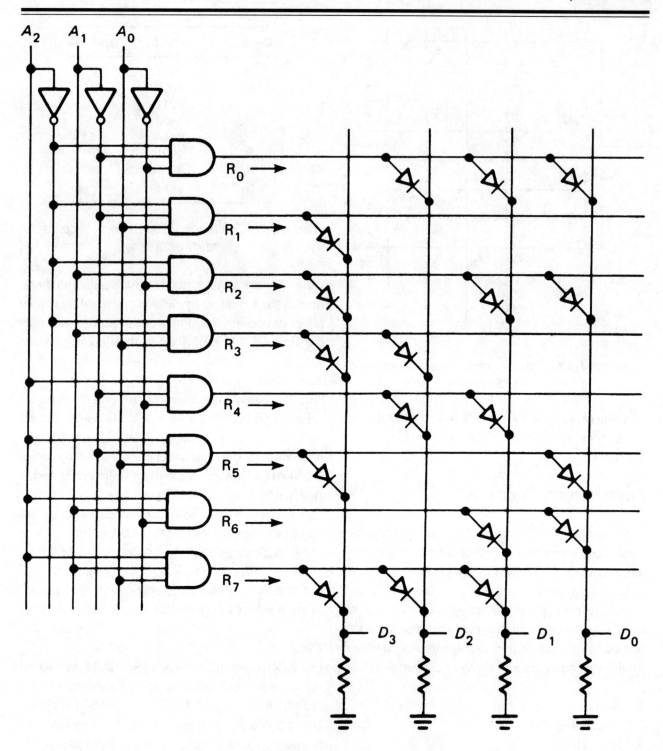

Figure 4-64. *An IC ROM with a decoder built-in.*

Programming

Figure 4-64 also shows where logic ones occur within the memory. Note at each logic one intersection there is also a diode. For example, register R_7 has diodes at D_3, D_2 and D_1 but no diode at D_0; therefore, the word is 1110. Keep in mind this diode occurrence coinciding with a logic one principle and refer back to *Figure 4-64* again. Here, AND gates and inverters comprise a decoder which eliminates the need for a switch for each memory location. A ROM's fixed pattern of

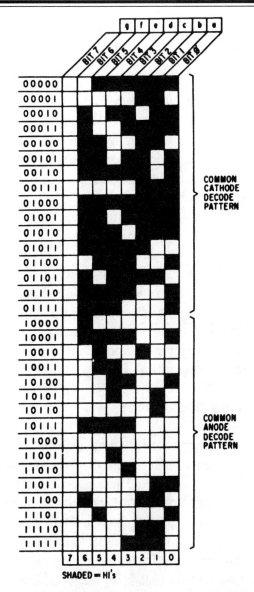

Figure 4-65. *A portion of an EPROM's memory representing the patterns for hex 0 to F for both common anode and cathode 7-segment LED displays.*

words is unalterable and corresponds to each unique address. In contrast, RAMs have transistors in place of diodes. This allows you to change or "write" data into their cells.

Memory Types

A memory electronically stores data in a binary format of logic ones and zeroes. Volatile and nonvolatile types refer to the memory's ability to retain data after removing power. A volatile memory loses its data in the absence of power. A nonvolatile memory retains its data in the absence of power.

The next major distinction is if the memory is a RAM (Random Access Memory) or a form of ROM (Read Only Memory). A RAM allows writing data to it, and a ROM, as its name implies, allows you to just read only data from it. RAMs come in 8, 16 or sometimes even wider configurations. There are two types of RAM, static and dynamic. A static RAM uses MOS flip-flops and stores data indefinitely, as long as power exists. A dynamic RAM contains capacitors and MOSFET transistors. Since a capacitor's charge eventually leaks off, you must refresh this stored data. Dynamic RAMs use a separate clock for refresh pulses every few milliseconds. A dynamic RAM's duty cycle, or ratio of ON time to OFF time, is approximately 1% or less; therefore, volatile RAMs are very power efficient.

Types of ROM

There are several types of ROMs, the most popular of which are:

1. The factory programmed mass produced ROMs based upon a pattern you specify in advance, permanently etch its substrate mask pattern.
2. The EPROM or electrically programmable ROM.
3. The EEPROM or electrically erasable PROM.

Figure 4-58 showed a multiplexing arrangement using latches. You could also use EPROM or EEPROM memories for a more complex 7-segment LED display multiplexing design in the development stages to verify your concept, or in products manufactured in small quantities. You can program a pattern to display hex outputs from the four data inputs, A to D. (See *Figure 4-65*) Note this accommodates both common anode and common cathode displays. If you can, photocopy *Figure 4-65* and enlarge it to about 200%. Then fold it exactly at the common anode and common cathode display's intersection, or paper's crease. You will observe they match with all black squares overlapping other black squares and all open squares overlapping open squares. This mirror image, before folding, shows how both the common anode and common cathode displays are logic complements of each other. This requires a paltry 32 bytes of memory. Horizontal rows in *Figure 4-65* specify words and you'd use a 5-bit word from 00000 to 11111. This equates to 32 positions or 16 combinations for both common anode and common cathode displays. This is 0 to F in hex.

An EPROM's Optical Properties

An EPROM has become a less popular memory technology, but it still is viable in many development applications. An EEPROM though has replaced it in many applications. An EEPROM allows you to electrically write in a pattern, and read it back. If you wish to alter this pattern, you may do so by electrically erasing it without removing it from the circuit.

The circa 1992 AMD (Advanced Micro Devices) EPROMs use a 253.7 nm UV-C exposure lamp within one inch of their surface with an intensity of 12,000 μW/cm^2. This accomplishes complete erasure in 15-20 minutes. Exposure to sunlight and/or fluorescent lights eventually erase an EPROM. Therefore, if you use an EPROM outside, or in an environment where they might encounter these light sources, place a piece of opaque tape over their exposure window. Any light source below 400 nm will erase an EPROM, but takes longer. As an example, barber sterilizing equipment uses shorter wavelength UV lamps which are more energetic, and not surprisingly, have greater potential for biological damage. These too, though, take longer to erase EPROMs.

Tri-State™ Memories

Figure 4-66 is a three-state static ROM. The third state is a high impedance state equivalent to an open circuit. This is useful in systems using the bus concept in which you have to electronically isolate various portions of a circuit. The TSL230 intelligent optoelectronics IC in Chapter 17 has an O/E pin which places this IC is the tri-state™ mode.

Fundamentals of 5 x 7 Dot Matrix LED Displays

Unlike 7-segment displays, a 5 x 7 dot matrix display does not come in either common cathode or common anode configurations. A 5 x 7 dot matrix display only connects it anodes and cathodes together on an individual row and column basis (not the whole display). As their name intuitively suggests, a 5 x 7 dot matrix optoelectronic display is a physical grouping of 35 individual LEDs occurring in a 5 x 7 column-to-row format, refer back to *Figure 4-61*. This minimizes the total number of external pins to just 12 (one pin each for 7 columns and 5 rows). It also greatly simplifies addressing, decoding and driving. This is despite these displays containing 35 individually accessible (addressable) LEDs! The intersection of each LED at each row and column represents one of these individually accessible (addressable) LEDs.

Physical Appearances

Figure 4-61 showed a 5 x 7 dot matrix display with rather pronounced half-sphere shaped divots resembling a golf ball's surface. These occur exactly on top of each LED dot. *Figure 4-62* showed a smaller multiple 4 character 5 x 7 dot matrix LED display. These types of displays are indistinguishable, if you don't look at their part numbers, just like identical styles of common cathode and anode 7-segment displays. Both types of these 5 x 7 dot matrix displays also have all their 12 pins in an identical pin-out order, and serve identical purposes. To troubleshoot circuits using these displays, especially in their breadboard stages, it would be prudent to manually illuminate just a single LED, based on what you believe is the display's schematic. If it fails to illuminate, it is highly likely you have the wrong internal, electrically-connected type display. This is very understandable under these circumstances!

There are single digit 5 x 7 dot matrix displays with their 35 LED dots composed of various combinations of multiple colors though, which are visually outstanding.

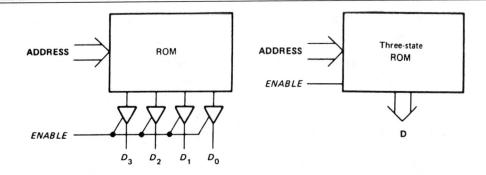

Figure 4-66. *A three-state or "tri-state^{tm}" logic type of memory.*

Types of 5 x 7 Dot Matrix Displays

There are two types of electrically connected 5 x 7 dot matrix LED displays regarding the inter-display electrical bussing of their rows and columns. These displays have either:

1. Rows of electrically common anodes and columns of electrically common cathodes.
2. Columns of electrically common anodes and rows of electrically common cathodes, referred to hereafter as Type 1 and Type 2 respectively.

Memory Capacities and Types of ASCII Character Sets

Memories allow flexibility to completely form many characters and symbols. (Refer back to *Figure 4-59.*) This lists the limited 64 and full 128 character ASCII set. The full 128-character ASCII set is twice as wide, but equal in height, to the limited 64-character ASCII set which is the chart's "inner" or middle eight columns. The full set requires 1K of memory, half the capacity of either our RAM or EPROM. A limited 64 character ASCII set requires just 1/2K or 512 bytes. So in our 2K memories you'd have plenty of space to forms customized characters, such as forming arrows pointing in both horizontal and vertical directions. Most designers closely adhere to this set; however, some exercise their latitude to replace some standard characters with their own characters or symbols.

Contrasting the Two Different Types of Memory-Based Drivers

The RAM/EPROM decoder/driver experiment board has both an EPROM and a RAM memory to decode and drive display types 1 and 2. You will use the RAM in both the PROGRAM and RUN states since you'll program its data. However, you can only read from an EPROM so it only uses the RUN position on jumper JP3 in *Figure 4-67.* You have to program the 2716 EPROM with an external EPROM programmer. Remember its lighting constraints so you may have to place a piece of opaque over its programming window. Since each memory based portion of this board drives a different type 5 x 7 dot matrix display, they are, in that regard, mirror images of each other. Both memories program 35 bits, one for each LED dot, and do so with an 8 x 8 or 64-bit inter-memory section. This leaves unused bits which cause three "phantom" columns and one "phantom" row (to be explained).

Generating 5 x 7 dot matrix LED alphanumeric characters uses power strobing **and** scanning. Electrically, horizontal scanning sequentially energizes one column at a time, from left to right. Vertically scanning of the rows does not occur since each memory type produces its 8-bit word in one step, not in a sequence. Your RAM/EPROM dot matrix driver experimentation board purposely uses each type of 5 x 7 dot matrix LED display. You may recall our previous 7-segment LED display multiplexing example used data latches. This board uses both data latches and two different types of memories. The **EPROM** memory-based circuit exclusively drives dot matrix display type 1. The **RAM** memory-based circuit exclusively drives dot matrix display type 2. (See the displays in *Figure 4-67*'s right side.)

Understanding the EPROM-Based Driver

If you place the 3-pin jumper, JP3 in *Figure 4-67*, into the **RUN** position, the 555 timer, U11, connects a clock to the 74LS161 counter, U10. The counter outputs (QA, QB and QC) count from 0 to 7, for a total of eight counts. When the count reaches 7, it recycles back to 0. This recirculating 8-count counter's outputs connect to the lowest address lines of U6, the 2716 EPROM, and to the 6116 RAM, U8. The outputs also connect U12, the 74LS138 data selector's select lines. When the counter counts from 0 to 7, the outputs of the 74LS138, Y0 through Y7, activate in sequence. Please note the first count is zero to correspond with that address and memory location. These outputs connect to open collector inverting buffer drivers within IC U3. IC U1's first five open collector inverting buffers (U1A through U1E) turn on the NPN transistors connected to the columns of the first 5 x 7 dot matrix LED display. Inverting buffers U1F, and all of U2, drive the rows of the first display in the upper right corner of *Figure 4-67*.

You may use a 2N3904 for the 10 NPN transistors, Q1 through Q10. However, if you can find one, a transistor array with five NPN transistors with all common emitters and another five transistors with all common collectors, or some combinations thereof, would be ideal and greatly reduce parts count to conserve board space.

When the counter's count is 0, Y0 of the 74LS138 goes low. This low Y0 connects to buffer U1A. When its U1A's input is low, its inverted output naturally is high. This turns on transistor Q1 and supplies power to the left column of the dot matrix display. If U1F has a low output, current flows through the upper left LED, series current limiting resistor R6, and inverting buffer U1F. This illuminates the upper left LED. If U1F and

buffers U2A through U2F are all low, this illuminates all LEDS in the left column.

The following discussion, when referencing data in memories, uses the customary convention of referring to their contents in the hexadecimal numbering base, not binary! As the count progresses from 0 to 4, transistors Q1 through Q5 supply power to the columns in a left to right **horizontal** scanning sequence. Activating each column changes the 2716 EPROM's address lines. With each address line change, the 2716 EPROM's data also naturally changes. This data drives buffers U1F and all buffers of U2 (U2A through U2F). You only use data bits 0 through 7. These data bits change the row LEDs, which light if the character has this now/column intersection, when each column activates. Stated differently, if a data bit from the 2716 is high, the inverted output from the buffers is low, and the row LED for the activated column lights.

To display the letter "T" in the 5 x 7 dot matrix LED display you must strictly adhere to the following EPROM sequence:

1. Program the data bits of EPROM address 0, with hexadecimal 1.
2. Program the data bits of EPROM address 1, with hexadecimal 1.
3. Program the data bits of EPROM address 2, with hexadecimal FF.
4. Program the data bits of EPROM address 3, with hexadecimal 1.
5. Program the data bits of EPROM address 4, with hexadecimal 1.

To display "T," the memory must allow count 0 to supply power to the left column. You may wish to simultaneously refer to the immediately preceding 5-step procedure again. Address 0 only supplies the data bits 1 hexadecimal or 0001 (binary) to the first (least significant) bit of the EPROM address 0, in the order: low, low, low and then high (Ø, Ø, Ø, 1). Inverting buffer U1F inverts this high to a low. This lights the top LED in the left column. Count 1 supplies power to the next column via transistor Q2. Address 2 again supplies the data bits 1 hexadecimal (0001 binary).

This lights the top LED of column 2. Count 3 powers the center column's top LED through Q3. Address 2 supplies data bits FF hexadecimal. All data bits are high, and the inverters U1F and U2A through U2F are low. This lights all the LEDS in the center column.

You do not use the next three address bits. These are not connected, but if there were eight columns, instead of five, they would control the rows for these other three nonexistent "phantom" columns. Only the five counts 0 through 4 power the dot matrix display columns. Counts 5 through 7 do not supply power to any column on the dot matrix display, obviously since none of these columns exist! This means the display is actually only on for 5/8 or 62.5% of the time.

Count 3 powers (actually supplies a path to ground for) the next column through transistor Q4. Address 3 supplies data bits 1 hexadecimal. This again lights the top LED in that column. Count 4 supplies power to the right column through transistor Q5. Address 4 supplies data bits 1 hexadecimal, again, lighting the top LED in the right column. Each of these power strobing events occur so fast your eye perceives the dot matrix display as a continuously powered VLS.

To display a different character, you may select address lines A3 through A10 of the 2716 EPROM. Each character requires 8 bytes of 8 bits each for 64 bits. The 2716 EPROM is a 2K EPROM. This means it stores 2048 bits. Since each displayed character uses 64 bits, the 2716 EPROM has the capacity to store 2048/64 or 256 characters. The following two equations quantitatively define this; although, you rarely define memory capacities in bits:

**Memory capacity (expressed in bits)/64
= the number of possible characters**

**Memory capacity (expressed in bytes)/8
= the number of possible characters**

The first character occupies the address space 0 through 7 in EPROM. The next character occupies address space 8 through 15 in EPROM. The third character occupies address space 16 through 23 etc.

You can access each EPROM character with a binary sequence from 0 through FF hexadecimal (256 in decimal) on its address lines, A3 through A10. This provides 2^8 or 256 possible combinations of characters or symbols.

The EPROM actually has no program to direct circuitry to do anything! It just contains a succession of continuous or adjacent bytes, starting at hex address 00, and progresses seven more memory addresses to form each 5 x 7 dot matrix character or symbol. This is true for both types of on-board memories. With a full 128-character ASCII set, the addresses start at 00 hex and progress to 128 • 8 = 1024 in decimal, minus 1, or 1023 in decimal. We subtract 1 because location 00 contains data, and is our starting point. The decimal number 1023 is 3FF in hex. This is the address of the last entry in the full 128-character ASCII set. The binary equivalent of 3FF hex is: 11 1111 1111. This is why you need 10 address lines to accommodate 3FF.

Programming the RAM-Controlled Dot Matrix Driver

To program the RAM display driver, you must thoroughly understand how the RAM drives this type dot matrix display. It operates in much the same way as the EPROM circuit, with two exceptions:

1. It is volatile so you must program it each time the power is turned off.
2. It drives a 5 x 7 dot matrix LED display (with the other configuration of internally connected cathodes and anodes) in just the opposite manner the 2716 EPROM did.

The RAM causes the sourcing (supplying) of current and the EPROM causes sinking of current. When you drive a 5 x 7 dot matrix LED display with the anodes connected to the columns, you must supply voltage to the columns, and a low resistance path to ground for the rows. Conversely, when you drive a 5 x 7 dot matrix LED display with the cathodes connected to the columns, you must supply voltage to the

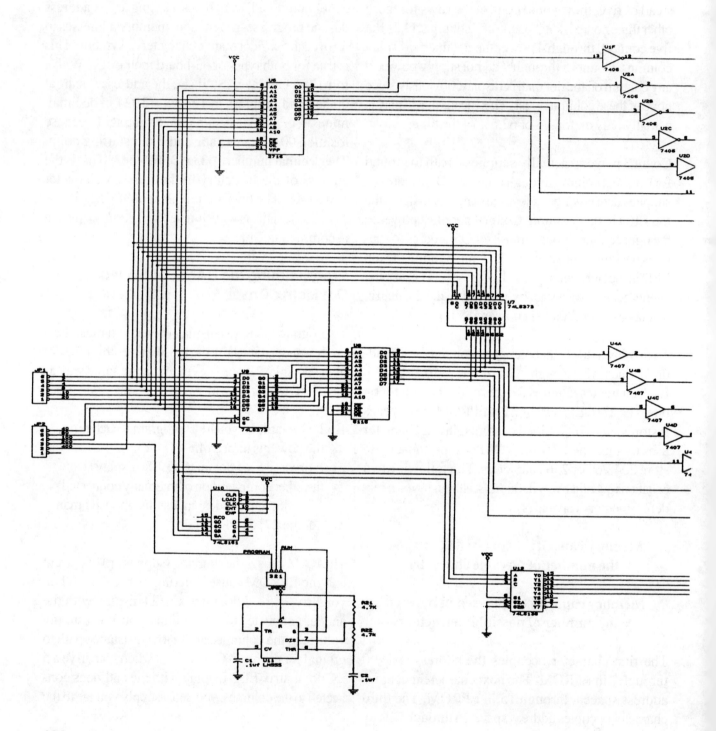

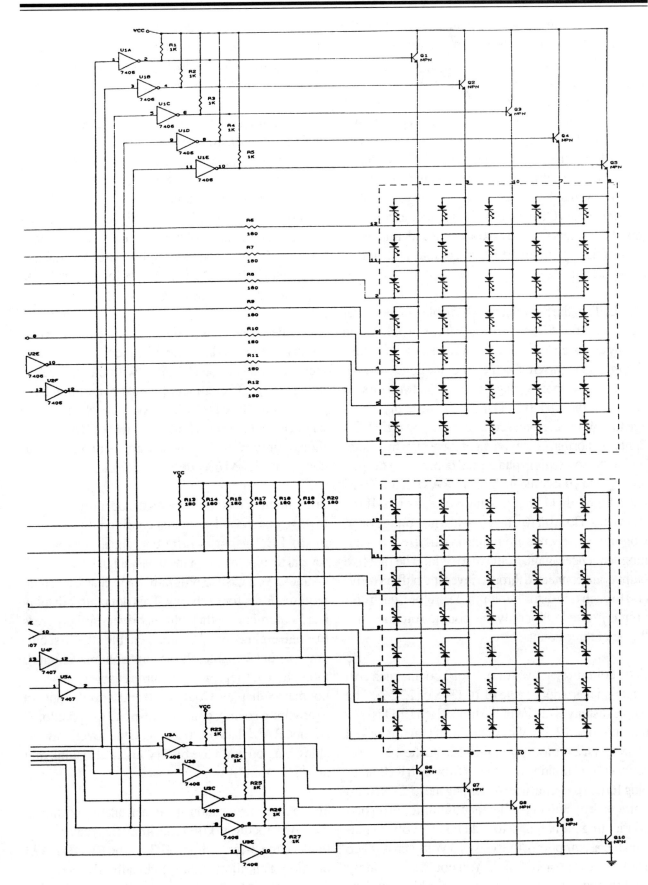

Figure 4-67. The RAM/EPROM decoder/driver schematic.

rows, and a low resistance path to ground for the columns. You can use the BCG board for this.

Stated differently, operation of this RAM's 5 x 7 dot display requires power (a +5 VDC potential) on each row and grounding of each column, or tying it low. The 2716 EPROM display decoder/driver section supplies (draws in) current to power to the columns so the buffers can "pull" their **rows** low. In contrast, the 6116 RAM display buffers supply power to each row, and transistors Q6 through Q10, to pull the **columns** low. Transistors Q6 and Q10 sequence identically to Q1 through Q5 in the EPROM dot matrix decoder/ driver design.

The LED in the 5 x 7 dot matrix display in the lower right corner of *Figure 4-67* only turns on when you supply a voltage to its row, and ground its column line. For instance, moving to the other diagonal display corner, if you apply voltage to row 1, and you ground column 1, the top left LED lights. If you apply a voltage to row 1 and ground column 5, the top right LED lights. In normal operation, lighting occurs from supplying the rows with voltage, and depends on the pattern of data you stored in the 6116 RAM. If the RAM has a high bit (1), that row has voltage on it. If the RAM has a low bit (0), that row has no voltage on it. Columns experience grounding in a horizontal scanning sequence, again, from left to right. The first five outputs are connected through inverting buffers (drivers) to each transistor's column. Therefore, each column is grounded for only one count. You don't use the other three counts.

Each column is active for only one count, each column and has a duty cycle of 100%/8 or 12.5%. The whole display is on 5/8 of the time. Each LED is on a maximum of 12.5%, if it occurs in all your displayed characters, a remote possibility! Note the 12 180W series current limiting resistors, R6 through R20, skipping R16 in the sequence. Assuming the LED forward voltage drop and the collector-to-emitter drop of a conducting NPN transistor add up to 2 volts (a reasonable assumption), this leaves 3 volts across each 180Ω resistor for 16.7 mA. You could quite safely slightly reduce these resistor values. This would in-

crease brightness, without damaging the dot matrix display's internal LEDs. The present values are conservative with a 12.5% LED power strobing rate, and a 12.5%, or usually less, typical LED activation rate.

The preceding procedure to form the letter "T" clarifies this. Its constituent ON dots, corresponding to the display's five columns, were (expressed in hex) 01, 01, FF, 01, and 01. Therefore, 1, 1, 5, 1, and 1 dots were ON within this 5 x 7 display, capable of lighting all of its 35 LED "dots." Since each column is only energized 12.5% of the time, and in this example, only 9/35ths of the possible lights were ON. Each of the display's 35 LEDS averaged being on 12.5% • 9/ 35, or just 3.2%, of the time.

Since the counter has a count of eight, and the RAM again has an 8-bit data path, like the EPROM, it creates an 8 x 8 section (64 bits), with the dot matrix using 5 by 7 (35 elements), or just part of that 8 x 8 array within the 6116 RAM. Again, each displayed character again uses 64 bits (8 bytes) of RAM, even though only 35 bits are necessary, per character, to control this RAM decoder/driver display.

Programming this RAM based decoder/driver circuit to drive this second particular configuration 5 x 7 dot matrix LED display is different. The driver requires several manipulations of the binary input lines for each character you wish to program. Each character again requires 7 bits for each row. This dot matrix display has 5 columns, but the column drivers still progress through eight counts. Count 0 enables the first column. Count 1 enables the second column. Counts 0 through 4 each represent a column enable in this 5 x 7 dot matrix display. Counts 5, 6 and 7 don't connect to anything so they again correspond to "phantom" counts. RAM data bits 0 to 7 feed the seven row enables, but you don't use bit 7 (you can also think of it as a "phantom" row).

To program one location, the first available location in the RAM, (address 00 hexadecimal) first requires you to place the 3-pin jumper, JP3, in the **PROGRAM** position. This allows you to manually advance or increment the column counter from 0 through 7, rather than allowing the counter to automatically advance it. The programming procedure this time is as follows:

1. First, place address 00 hexadecimal into the RAM by setting binary inputs 1 - 80 to 0 (low voltage).

2. Set binary bit 100 low to latch the address into U7, the 74LS373 address latch.

3. Toggle the binary bit 200 from high to low until the column counter is on the first column.

4. Set the binary bit 400 low. This enables the data latch.

A COMPARISON of the RAM/EPROM MEMORY-BASED DOT MATRIX DRIVE CIRCUITS

	RAM	EPROM
This type memory-based circuit drives 5 x 7 dot matrix displays with common anode rows and common cathode columns.	NO	YES
This type memory-based circuit drives 5 x 7 dot matrix displays with common cathode rows and common anode columns.	YES	NO
This type memory of the memory-based circuits has more than sufficient capacity to display all 128 5 x 7 dot matrix ASCII characters.	YES	YES
You can program the contents of this type memory from on board this Experimentation board.	YES	NO
You program it with the Jumper, JP3, in the Program position and mode.	YES	NO
This type memory-based circuit's memory uses 64 inter-memory bits for each 35-bit 5 x 7 dot matrix character or symbol.	YES	YES
This type memory-based circuit to drive 5 x 7 dot matrix displays horizontally scans its columns from left to right, and activates them in sequence.	YES	YES
Each "dot" or LED within each 5 x 7 display's columns in this type memory-based circuit is only ON for 12.5%, or usually less.	YES	YES
This type memory-based circuit has one "phantom" row and three "phantom" columns.	YES	YES
You can access each character or symbol of this type memory-based circuit by using 0 to FF in hexidecimal on its address lines, A3 to A10.	YES	YES
This memory-based circuit loses its contents if its power is interrupted and/or shut off.	YES	NO
Each activated ON LED within this type memory-based circuit's displays draws 16.7 mA.	YES	YES
This type memory-based circuit sequentially activates its columns, from left to right, by sourcing current.	NO	YES
This type memory-based circuit sequentially activates its columns, from left to right, by sinking current.	YES	NO

Table 4-8. *The similarities and differences between the two memory-based circuits comprising the RAM/EPROM decoder/driver board.*

5. Set binary bits 0 - 80 so they light the row LEDS you want to program. **Note:** When you find the correct configuration, set binary bit 400 high. This writes the row data into the correct RAM location for column 1 of the first character.

6. Toggle the column counter high and low one time to increment (advance) the column counter to the next column (column 2).

7. Again, set binary bit 400 low and set the row information into binary bits 0 - 80.

8. When the correct LED pattern for column 2 is displayed, set binary bit 400 high to write data into the RAM.

9. Repeat this procedure for each of the seven columns.

You can program additional characters by setting binary bit 100 high and placing a new address, (such as 01 hexadecimal), on binary lines 0 - 80. Set binary bit 100 low to latch the new address and repeat the same preceding steps for the next character. When all the characters you desired are programmed into the RAM locations, the following procedure displays them:

1. Set all binary inputs high.

2. Connect jumper JP3 to the **RUN** position to enable the automatic column counter.

3. Set the address of the first character to be displayed into binary bits 0 - 80.

4. If you enter address 00, the first character programmed will be displayed.

5. If you enter 09 hexadecimal, you will display the 10th character.

Table 4-8 shows the similarities and differences between the two memory-based circuits comprising the RAM/EPROM decoder/driver board.

Chapter 4 Quiz

1. Which type LED produces the greatest light output and casts light in the narrowest pattern?
   A. Diffused
   B. Non-diffused
   C. Tinted
   D. Water clear

2. Which type LED has tiny glass particles in its plastic lens?
   A. Diffused
   B. Non-diffused
   C. Tinted
   D. Water clear

3. Which type LED has the same color in its ON and OFF states?
   A. Diffused
   B. Non-diffused
   C. Tinted
   D. Water clear

4. Which type LEDs are primarily made for backlighting?
   A. Diffused
   B. Non-diffused
   C. Tinted
   D. Water clear

5. The ClipLite™ corrects which inherent LED problem?
   A. A small viewing angle
   B. Shifts in color as an LED ages
   C. Guarding against ESD
   D. Both A and C

6. Silver sulfide is better known as?
   A. Oxidation
   B. Tarnish
   C. Rust
   D. None of the above

7. Storing silver lead LEDs in a plastic bag with what creates a vapor pressure, which keeps (excludes) free air out:
   A. Lemon juice
   B. Acetic acid
   C. Petroleum napthelene mothballs
   D. None of the above

8. More advanced tarnishes require:
   A. Stronger more active fluxes
   B. Weaker less active fluxes
   C. An acetic and sulfuric acid mixture
   D. Both A and C

9. The experimenter's BCG PC board produces _____ outputs.
    A. Eight
    B. Four
    C. Twelve
    D. Sixteen

10. Dual LED manufacturers strive for?
    A. Closely matched light intensity outputs.
    B. To exclude the color red.
    C. To exclude the color green.
    D. None of the above.

11. Your eye perceives a continuous color when switching dual color LEDs at or above what rate?
    A. 1.6 Hz
    B. 22 to 30 Hz
    C. 160 Hz
    D. 1 Hz

12. Color mixing in a dual color LED involves:
    A. Forward biasing one color LED and then suddenly forward biasing the other color LED within the dual LED package.
    B. Applying a pseudorandom number generator to the dual LED.
    C. Disabling one LED permanently while manipulating the other.
    D. None of the above

13. You increase the color mixing resolution of a dual LED's colors by:
    A. Allowing fewer steps in random duty cycle variations.
    B. Allowing more steps in random duty cycle variations.
    C. Allowing more steps in continuous duty cycle variations in one direction followed by the other.
    D. None of the above.

14. Power strobing a dual color LED entails:
    A. Pulsing the dual LED with random duty cycle variations.
    B. Pulsing the dual LED with short duration bursts of current, often exceeding the LED's continuous (DC) current rating.
    C. Both A and B
    D. None of the above

15. Using high quality expensive optical glass filters is one way to:
    A. Increase an LED's intensity, as observed by a viewer:
    B. Save power.
    C. Increase the LED's life.
    D. None of the above.

16. Incandescent bulbs have virtually remained unchanged since who invented them?
    A. Voltaire
    B. Volta
    C. Hertz
    D. Edison

17. Neon lamps have what kind of glass envelopes?
    A. Standard glass
    B. Leaded glass
    C. UV sensitive glass
    D. IR sensitive glass
18. A neon lamp's excitation voltage is:
    A. The voltage at which gas ionization breakdown occurs.
    B. The voltage which damages the neon lamp.
    C. The voltage at which neon lamps normally run.
    D. The voltage below which gas ionization and current flow cease.
19. One of the most admirable qualities of a neon bulb is:
    A. It has a longer life than an LED.
    B. It has more light output per applied power than an LED.
    C. It can absorb voltage surges and transients, without damage.
    D. All the above.
20. A halogen lamp:
    A. Actually only has small traces of a halogen element.
    B. Allows its ionized gas to circulate back to its filament, which can reach 2,500 °C.
    C. Is ideal for a spectroradiometer.
    D. All of the above.
21. Your eye most readily perceives which color as the brightest in LEDs?
    A. Orange
    B. Yellow
    C. Green
    D. Red
22. A duty cycle is composed of a signal's on time and ___?
    A. Its period
    B. Its frequency
    C. Its off time
    D. None of the above
23. A moving point or dot LED indicator display driving circuit allows ___?
    A. All LEDs to be on at once
    B. Just the bottom LED to be on
    C. Just the top LED to be on
    D. Just one LED to be on, which indicates the state of the input in proportion to the circuit's total input range.
24. The serially connected string of equal value resistors in the op amp bar graph driver circuits serves what purpose?
    A. It is a voltage divider which equally divides supply voltages into increments equal to the number of resistors you use, minus one.
    B. A voltage divider which equally divides supply voltages into increments equal to the number of resistors you use.
    C. It divides the supply voltage in half.
    D. None of the above.

25. How does the discrete transistor and current steering diode driver for series stacked LEDs relate to Equations 4-15 and 4-16?
    A. It conforms to the equation for a straight line.
    B. It conforms to a parabola.
    C. It conforms to a hyperbola.
    D. None of the above.
26. The LM3916 bar graph driver IC:
    A. Is designed to have a linear output.
    B. Is designed to have a logarithmic output.
    C. Is designed to have a logarithmic output referenced to 600 Ohms to indicate (vu) or volume units for audio applications.
    D. Both B and C.
27. The LM3914:
    A. Is designed to have a linear output.
    B. Is designed to have a logarithmic output.
    C Is designed to have a logarithmic output referenced to 600 Ohms to indicate (vu) or volume units for audio applications.
    D. Both B and C.
28. The LM3915:
    A. Is designed to have a linear output.
    B. Is designed to have a logarithmic output.
    C Is designed to have a logarithmic output referenced to 600 Ohms to indicate (vu) or volume units for audio applications.
    D. Both B and C.
29. Having an output referenced to 600 Ohms describes (vu) or audio volume units, T or F?
30. The LM3914 series driver ICs have both a dimness and alarm function, T or F?
31. The bar graph driver display application producing, an exclamation point output, tends to make viewers quickly lose attention, T or F?
32. You realize the indicator and alarm functions on these LM3914 series of bar graph driving ICs by switching between their dot and bar modes, T or F?
33. Most commercially available displays have slightly bowed edges making it impossible to stack them, T or F?
34. A 7-segment display represents its segments by the letters "a" through "h", T or F?
35 Displays differ by their internal connections. They are either internally connected in common anodes or common rows, T or F?
36. The 7447 7-segment decoder driver IC is for common cathode displays, T or F?
37. The 7448, nearly an identical twin to the 7447 IC, is for driving common anode displays, T or F?
38. In one experiment, the monostable or one-shot oscillator substitutes what for its brightness control pot?
    A. A fixed value resistor.
    B. Two resistors in parallel.
    C. A small resistor-capacitor combination.
    D. A photocell whose resistance is light dependent.

39. Two annoying problems of driving 7-segment displays, with the 7447, and many other similar ICs are?
    A. No lamp test and ripple blanking
    B. No ripple blanking in and ripple blanking out
    C. No ways to form hexadecimal numbers and completely form the numerals 6 and 9.
    D. None of the above.

40. A 5 x 7 dot matrix display gives you the ability to form unique character sets and symbols, T or F?

41. Generating 5 x 7 dot matrix characters or symbols uses LED power strobing and the scanning technique, T or F?

42. One simple way to form a hex number "A" involves using what type of display, along with the 7447?
    A. A common cathode
    B. A common anode
    C. A common cathode display with a left hand decimal point
    D. A common anode display with a right hand decimal point

43. A kilobyte of memory represents how many bytes?
    A. 1,000
    B. 1,024
    C. 1,032
    D. 1,004

44. A byte has 4 bits, T or F?

45. A nibble has 8 bits, T or F?

46. Nonvolatile RAMs retain their contents after you shut off power, T or F?

47. A dynamic RAM uses a lower duty cycle strobing method and, along with its MOS flip-flops and capacitors, cause it to be an inefficient power consuming memory, T or F?

48. Static RAMs have MOS flip-flops and capacitors, T or F?

49. A tri-state™ memory's third state represents a high impedance state, which is ideal for bus oriented systems, T or F?

50. You erase EEPROMs with UV light, T or F?

51. The LM3914 series of bar graph driver ICs eliminate the need for:
    A. Current limiting output resistors for its driven LEDs.
    B. Precision voltage divider resistors on its inputs.
    C. The use of external light sensors, such as photocells, to control LED brightness.
    D. A and B

52. Why is it a concern for the series connected resistors in a voltage divider to have matching temperature coefficients?
    A. If they all match, the resistors will increase while the bottom resistors decrease in value as temperature rises.
    B. If they all match, they will all drift in unison in the same direction and with the same magnitudes of change.
    C. If they all match, they will experience small changes in resistance in proportion to changes in input voltage.
    D. None of the above.

53. The moving dot indicator, comprised of a quad op amp, has its LED current limiting resistors connected in parallel with +12 VDC, T or F?

54. The moving bar graph type of LED indicator, comprised of a quad op amp, has its LED current limiting resistors connected in series with +12 VDC, T or F?

55. The LM3914 IC has no reference adjustment control pin, T or F?

56. Common anode LED displays are more prevalent than common cathode LED displays, T or F?

57. A halogen is one of any of five elements, all of which reside in the VIA portion of the Periodic Table, T or F?

58. Both dot matrix and bar graph displays have multiple color elements, T or F?

59. The transistor and current steering diode circuit in Figure 4-31 for driving series arrayed LEDs, first lights the LED at the top of the "stack", T or F?

60. Nonlinearity in the LM3914 is typically 0.1% or better, T or F?

61. The square wave adjustable duty cycle op amp oscillator, U3 in Figure 4-31 uses:
    A. Just negative feedback
    B. Just positive feedback
    C. Both negative and negative feedback
    D. No feedback

62. The square wave adjustable duty cycle op amp oscillator, U3 in Figure 4-31, is composed of:
    A. Three smaller integral circuits
    B. An integrator and a latch only.
    C. An integrator, a comparator and a latch.
    D. Both A and C.

63. The square wave adjustable duty cycle op amp oscillator, U3 in Figure 4-31, has a pot, R39, which does what?
    A. It controls the duty cycle.
    B. It controls the output's polarity.
    C. It controls the amount of current the op amp draws.
    D. None of the above.

64. It is unnatural for an op amp:
    A. To be used as an oscillator.
    B. To be used as an amplifier.
    C. To have its noninverting (+) and inverting (-) input pins not have equal voltages.
    D. To be used with feedback.

65. If the pot, R39, in the square wave adjustable duty cycle op amp oscillator, U3 in Figure 4-31, had a small value compared to R35 and R38, and also had 10 turns, it:
    A. Would just provide a greater duty cycle span.
    B. Would just provide a smaller duty cycle span
    C. Would provide a smaller duty cycle span and much finer duty cycle adjustment resolution.
    D. Both B and C.

66. The RAM/EPROM decoder/driver board for 5 x 7 dot matrix LED displays has two displays on board which:
    A. Are physically different in appearance but electrically configured identically.
    B. Are identical physically in appearance but electrically configured differently.
    C. Are physically different in both their appearance and in their electrical configurations.
    D. Are physically identical in both their appearance and electrical configurations.

67. The two memories on the RAM/EPROM decoder/driver board with two 5 x 7 dot matrix LED displays:

   A. Have an EPROM with capacity to display all 128 ASCII characters and symbols and a RAM also capable of displaying all 128 ASCII characters.

   B. Have an EPROM with capacity to display just 64 ASCII characters and symbols and a RAM capable of displaying all 128 ASCII characters.

   C. Have an EPROM with capacity to display all 128 ASCII characters and symbols and a RAM capable of displaying just 64 ASCII characters.

   D. Have an EPROM with capacity to display just 64 ASCII characters and symbols and a RAM also capable of displaying just 64 ASCII characters.

68. The power strobing technique used on the RAM/EPROM decoder/driver board for 5 x 7 dot matrix LED displays:

   A. Overstresses and overdrives the LEDs.

   B. Is designed in a conservative fashion.

   C. Could slightly increase the LEDs' current flow, without concern.

   D. Both B and C

Chapter 5
Photocells, Photodiodes and Photomultipliers

Chapter 5
Photocells, Photodiodes and Photomultipliers

This chapter examines four topics:

1. Innovations in photovoltaic cell based alternate energy.
2. Photoconductive cells, also called light dependent resistors, or just "photocells."
3. Photodiodes.
4. Photomultipliers.

Photocells economically sense light in many digital and analog designs. Sensing the presence or absence of light is a digital operation. Sensing light intensity is a photocell analog operation. Photocells are more sensitive than photodiodes to visible light, such as in ambient light or color registration applications. But photodiodes also sense light, switch very fast, are linear over several orders of magnitude of light intensity, but, unlike photocells, require external amplification.

Photosensor Theory

A photosensor is an electrical device which responds to a change in the intensity of the light striking it. There are four types of photosensors:

1. Photoemissive.
2. Photovoltaic.
3. Photoconductive junction.
4. Photoconductive bulk effect.

The photoemissive type measures light by the emission in a vacuum of one electron per photon falling upon a metal photo cathode. Photomultiplier tubes use this technology of successive stages with secondary emission to amplify minute electron current. The photovoltaic type sensor generates a voltage across a P-N junction as a function of the

photons falling upon the P-N junction. This P-N junction is either silicon or selenium and requires no external power supply since it is self generating. The solar cell is an example of a photovoltaic sensor.

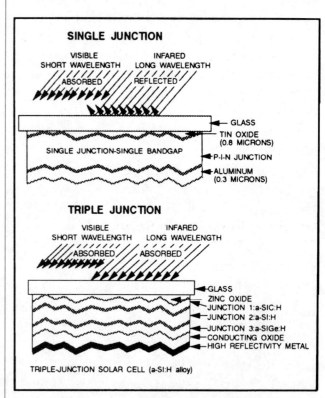

Figure 5-1. *Light trapping in silicon.*

Photovoltaic Based Solar Energy

Alternate energy, especially solar energy, was one of the highest interest topics within the scientific community the past 20 years. Numerous articles touted this promising new technology. However, the technology fell far short of any practical energy generation products. Granted, passive solar energy panels can heat your water heater, but any viable system must generate AC power, what industry and consumers use. Traditional fossil fuel and nuclear generated power is relatively cheap. Today though, alternate energy, solar, wind, geothermal and all others account for less than 1% of total U.S. energy generated.

Specially Made Photosensors

As an overview, before our detailed discussions of each, photovoltaic photocells differ from photodiodes in five ways:

1. A photocell's series resistance is about one ohm compared to a photodiode's 50 ohms.
2. A solarcell's depletion area is very narrow to provide a higher open circuit output voltage.
3. A photocell's sensing area is much larger than that of a photodiode.
4. A photocell's top layer (N type material) is very thin to allow its broad spectral response to even reach UV.
5. A photocell requires no external bias or voltage.

Advances in Materials Science

From the 1970s until about six years ago, research concentrated on increasing solar energy conversion rates of photovoltaic cells. This meant sophisticated and expensive silicon development. After attaining the state-of-the-art efficiency, increasing it just 1% doubled costs and another 1% increase in efficiency quadrupled costs, a point of obvious diminishing returns. Therefore, the Department of Energy's $55 million research project, PVM (Photovoltaic Manufacturing Technology), emphasizes less efficient photovoltaic cell technologies. These are at least 10 times cheaper to manufacture.

New materials science research focuses on thin-film devices, usually as narrow as one micron — in contrast to traditional solar cells as thick as 900 microns. Amorphous silicon does not have tightly packed very orderly lattice pattern arranged molecules. Its inherently disorderly structure is less energy efficient but much cheaper to produce, in contrast to semiconductor silicon, which is slowly grown and spun into ingots and later sliced into wafers. Energy Conversion Devices (ECD), refer to Appendix B for a listing of similar companies, has developed an 85-foot machine that coats sheets of stainless steel with thin film.

This inherent inefficiency is not casually accepted as an unfortunate trade-off. Ongoing thin film research addresses the faults of just capturing shorter wavelength colors (orange, green and blue) and not capturing longer wavelength colors (reds).

Solarex leads the triple cell or triple junction technology. (See *Figure 5-1*) This "trapping" technique harnesses red light's energy with triple layered cells. The layers use silicon-carbon, silicon alone, and silicon-germanium. This allows a 14% capture rate of total solar energy, with this energy cost predicted to plummet to 10 cents per kW/hr by the year 2000.

Astro Power uses crystalline silicon, rather than the amorphous type. The light trapping layers are between 50 to 100 microns thick enabling the building of giant solar cells, measuring 6 by 18 inches. These larger, plank-made cells avoid the traditional ingot process.

Concentrators

These use silicon, not in its traditional forms, to build large curved Fresnel (focusing) plastic lenses which direct and concentrate light. Concentrators offer 18% efficiencies but require sunlight bright enough to cast a shadow. Capturing the brightest

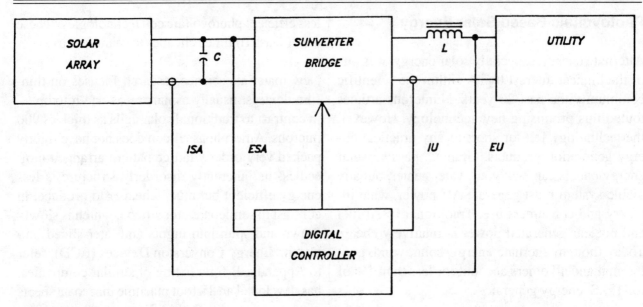

NOMENCLATURE:
ESA — SOLAR ARRAY VOLTAGE
ISA — SOLAR ARRAY CURRENT
EU — UTILITY VOLTAGE
IU — UTILITY CURRENT

Figure 5-2. *The Sunverter^tm's block diagram.*
(Courtesy of Abacus)

light of the day hitting a concentrator requires electronic circuits sensing luminous intensity by rotating and training the concentrator on the sun's brightest approach.

A Solar Array Controller

The Abacus Sunverter^tm advances the state-of-the-art in DC-to-AC inverters. These are one of a solar system's three system elements, along with solar arrays and the utility's interface. Two wires connect the solar array DC power to the Sunverter^tm input, and two wires connect its output to 240 VAC at the utility circuit breaker box of your home.

In the utility intertie operation mode, you need no storage device. The storage of fuel is retained at the utility generating station for the amount of power you provide by solar energy. For most U.S. utilities, the presence of sunlight coincides with peak load demands. Therefore, the solar energy supply of power amounts to "peak shaving," in utility jargon.

The Sunverter^tm has a microprocessor with software control for overall management and an *ASIC* (application specific integrated circuit), with hardware control, for microsecond response. *Figure 5-2* is the Sunverter^tm block diagram. The solar array delivers direct current to the Sunverter^tm bridge, which converts the DC to AC and delivers it to the utility. A capacitor filters the current drawn by the Sunverter^tm bridge, minimizing solar array ripple voltage. An inductor filters the pulse-width-modulated output voltage of the Sunverter^tm bridge to a level minimizing the inverter output current's total harmonic distortion.

The Sunverter^tm bridge has four *IGBTs* (insulated gate bipolar transistors) arranged in an "H" bridge with an isolation transformer connected across the center of the bridge. The IGBT is superior to the older bipolar transistors in response speed, safe operating voltage capability, efficiency, and has simplified snubber and clamp circuits. It requires two orders of magnitude less driver circuit power which reduces supporting parts' costs.

The digital controller (See *Figure 5-2* again) senses both input and output voltages and currents. A complex multi-loop control strategy provides on/off instructions to the Sunverter™ bridge's four output IGBTs. Accomplishing all control functions through a single power stage maximizes reliability, maintainability and efficiency, and minimizes cost, size and weight.

The digital controller performs the following:

1. Determines (at dawn) when solar array power is sufficient to deliver power to the utility.
2. Creates a sine wave reference for the inverter current.
3. Synchronizes the inverter current to the utility voltage.
4. Provides a bumpless turn-on when connecting to the utility.
5. Finds the maximum power point of the solar array.
6. Regulates the solar array voltage to retain operation at the maximum power point.
7. Provides shutdown protection for out-of-tolerance conditions at either the solar array or the utility interface.

The simplified block diagram does not show DC and AC circuit breakers, a static relay between the Sunverter™ bridge and the utility, and radio frequency interference (RFI) filters in both the DC and AC connections. The RFI filter at the DC input is vital since the solar array is a huge antenna at communication frequencies.

Maximum Power Tracking

A solar array produces a constant current proportional to the photon energy the array receives. This is "insolation." The array's open circuit voltage varies insolation in the solar array, and the solar array's temperature.

Figure 5-3 shows current-voltage curves of a 6 kW peak solar array with six levels of insolation and load resistances from zero (vertical axis, short cir-

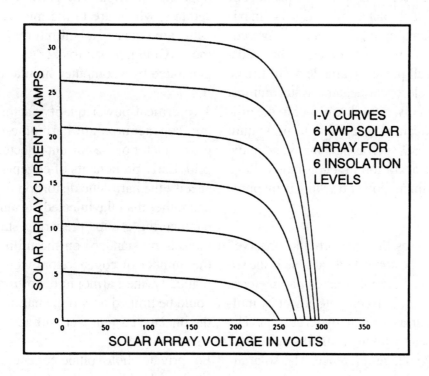

Figure 5-3. *Current vs. voltage for 6 kW solar arrays for six insolation levels.*
(Courtesy of Abacus)

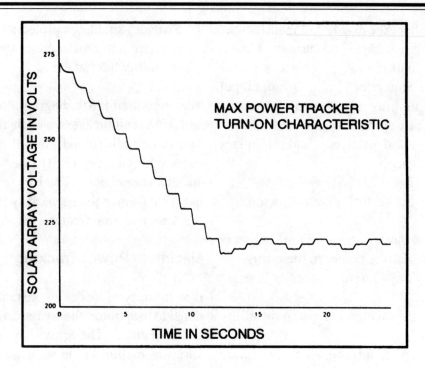

Figure 5-4. *Maximum power tracker turn-on characteristics.*
(Courtesy of Abacus)

cuit current) to infinity (horizontal axis, open circuit voltage.) These curves reveal a unique point on the knee of each curve at which the product of solar array current and solar array voltage is a maximum; this is the maximum power point. A prerequisite for maximum power tracking is the load's ability to absorb all power available at all times. Battery charges, electrical heaters, water pumps, and most significantly, returning power to the utility grid, are applications adaptable to maximum power tracking. Loads such as light bulbs and home appliances prefer operation in a constant voltage mode where maximum power tracking is inappropriate.

The Sunverter™ uses the "perturb and observe" tracking strategy. The controlled variable is the solar array voltage and the maximized parameter is the solar array power. You compute power by multiplying the solar array voltage by its current. The system is perturbed by taking a small step change in the solar array voltage reference. The solar array power is again computed, and the difference between the latest value of power and the preceding value compared. If the difference is positive,

the next step is in the same direction as the last. If the difference is negative, the next step is in the opposite direction. The perturb and observe procedures, which are in the microprocessor's software, are repeated ad infinitum. *Figure 5-4* shows the maximum power tracker turn-on characteristic generated by a computer simulation.

Two critical power quality factors for photovoltaic cogenerators are harmonic content and the power factor of the current injected into the utility grid. Early proponents of cogenerators wanted to specify the harmonic distortion of the utility voltage rather than the injected current. However, the engineering groups preparing standards realized there is no standard on the utility impedance. If the number of cogenerators were to expand dramatically, the harmonic distortion of the voltage could be limited only if the harmonic distortion of the injected current were limited.

For private photovoltaic cogenerator owners, one is the ideal power factor. Utilities owning cogenerators desire a power factor allowing their current entering the local distribution to be unity.

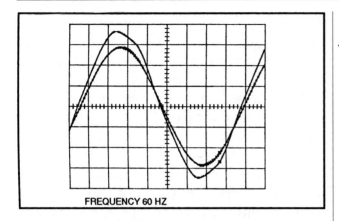

Figure 5-5. *Full load performance of the 5 kW Sunverter^tm; utility voltage (outer curve) 100V per division; inverter current (inner curve) 10 Amps per division. (Courtesy of Abacus)*

Both approaches minimize the owner's cost. The Sunverter^tm's power factor is set for unity, but virtually any power factor from zero lead to zero lag is possible. An early Sunverter^tm experiment proved 1 kW of solar array power can generate 10 k VAR at a zero leading power factor. For a utility this is equivalent to having a programmable power-factor-correcting capacitor bank.

Figure 5-5 shows the full load performance of a 5 kW Sunverter^tm. The utility voltage total harmonic distortion is 4.1%, but the photovoltaic cogenerator current being injected into the utility grid is 1.9%. A digital staircase reference sets the desired current value in a sampled data control loop independent of the utility voltage's wave shape. The phase-locked loop synchronizes the output current to the utility voltage per the zero crossing points of the phase reference's voltage.

The model 1992 photovoltaic cogenerator's cost is from 40 to 79 cents per watt. After the initial expenditure, the sun's radiant energy is free.

Photosensors

The two most popular photosensors are, first, photoconductive junction type sensors which include photodiodes and phototransistors. Next, the bulk effect photoconductors made from *cadmium sulfide* (CdS) and *cadmium selenium* (CdSe) are photoresistors or photocells. This device's current flow and exposure to bright light are proportional. *Figure 5-6* shows spectral responses of various photoconductors. Note CdS has about 2.4 eV while Si (silicon) has only 1.5 eV.

Photocells

Photocells made of CdS or CdSe have no junction, just uniform material upon which incident light induces a resistance change. They are truly light dependent variable resistors. A photocell's resistance is inversely proportional to the amount of light striking it. Increasing light intensity decreases resistance. The absolute value of resistance of a particular cell at a specific light level depends upon the photosensitive material used, the cell's actual physical size, electrode geometry, and the incident light's angle of arrival and spectral composition. Older type vacuum photocells were used for film sound pickup and in older photometers; however, their downfall was that their spectral responses greatly varied from unit to unit.

Selecting a photocell addresses two areas:

1. Optical characteristics.
2. Physical size and packaging.

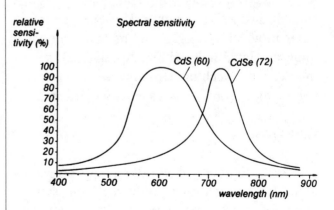

Figure 5-6. *Spectral responses or sensitivities of photoconductor materials used in photocells.*

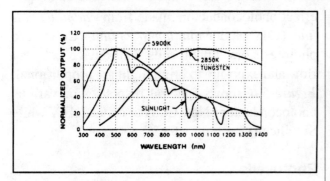

Figure 5-7. *The output of a blackbody source vs. wavelength.*

Each photocell has its own unique spectral response curve which indicates the portion of the light spectrum to which it is most responsive.

Your photocell design must use a photoconductive material and light source with maximum relative sensitivity. Relative sensitivity is the ratio of cell conductance obtainable, had the wavelength of the light source used been exactly equal to the peak wavelength of the material.

Color Temperature Response

Photocells are predominantly used in applications in which the light sensed is from one type of emitting source, such as sunlight, neon, incandescent, etc. This makes the overall response to a continuous light source very important. If a photocell has to operate over a wide color temperature range, it should have a flat response. However, there are cases in which a fluorescent ambient light has a tungsten signal light within. A CdSe cell solves this with its much higher response to the lower color temperatures of tungsten. (See *Figure 5-7*)

Sensitivity

This is the ratio of a photocell circuit's output divided by the light intensity of the input source. Obviously, in a simple circuit with a photocell, power supply and a relay all in series, the sensitivity of the photocell depends upon current flow which, in turn, is greatest for the photocell with

the lowest resistance. Using a photocell as the input to a high impedance FET has the voltage generated into almost an open circuit. Here, an ideal photocell is one with the ability to withstand large voltage changes. A high resistance photocell tends to impedance match this and operates best under these conditions.

A photocell used in a switching circuit should have the greatest slope of resistance. A photocell has a **light resistance** and a **dark resistance**. As their names imply, these are the resistances of the photocell in the light and in the dark. As a rule-of-thumb, the ratio of light-to-dark resistance in a photocell is 1,000:1. Resistances in bright light are in the 1 kilohm range, while in total darkness they are approximately 1 megohm. Clairex, a leading photocell manufacturer, has a range of light-to-dark resistances of 100:1 to 10,000:1. This ratio requires you to specify the elapsed time after you remove the light. This determines the **dark resistance**. Clairex specifies this as five seconds of dark, after exposure to 2 *fc* (footcandles).

A perfectly linear photoconductive material is one in which a given percentage change in light level results in the same percentage change in resistance over the entire range of illumination. No material is perfectly linear. However, Clairex's type 5H material exhibits a very predictable linearity. Generally photocells exhibit very linear relationships of light versus resistance in less than 1 *fc* of light. They become less linear above 1 *fc*. In the range of 10,000 *fc*, the photocell becomes asymptotic or resembles a capacitor's charging curve after five time constants (it approaches the maximum). When the photocell reaches this point, it is no longer useful.

Photocell Uses

Photocells in many packages sizes and shapes respond to very low light levels (moonlight) and very high light levels (direct sunlight), and have a wide dynamic range. That is their resistance with respect to applied light (no light vs. full light) ranges over

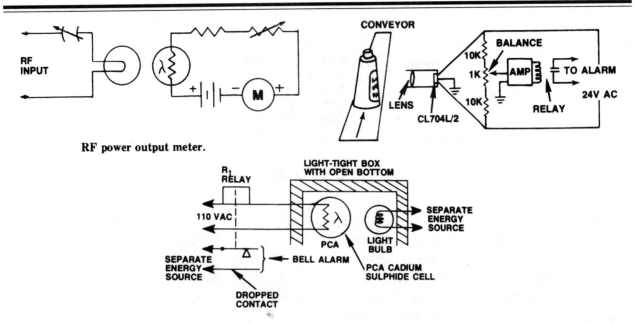

RF power output meter.

Simplified drawing of a smoke detector.

Figure 5-8. *Common photocell applications.*
(Courtesy of Clairex)

several orders of magnitude. Photocells are not bothered by noise or distortion, can operate from 50 to 400 VDC and on 120/240 VAC, and also are usable with almost any visible or near infrared source such as: LEDs, neon, fluorescent, incandescent, laser, sunlight and flame sources. *Figure 5-8* shows some common photocell applications.

Basic Theory

Semiconductor light detectors are:

1. Junction devices.
2. Bulk effect devices.

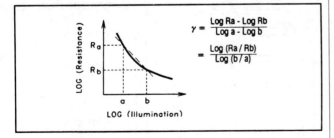

Figure 5-9. *A graph of a photocell's resistance vs. illumination.*

Junction devices in the photoconductive mode use the reverse characteristic of a P-N diode. Here, a P-N junction diode acts like a light controlled current source. The output is proportional to incident illumination and is relatively independent of applied voltage. Silicon photodiodes are examples of this.

Bulk effect photoconductors have no junction! *Figure 5-9* shows their resistance decreases with increasing illumination, allowing more current flow. This resistive characteristic gives bulk effect photoconductors a unique quality. You can vary the signal current from the detector over a wide range by adjusting the applied voltage. That is why they are also called light dependent resistors.

Photocells are thin film devices made by depositing a layer of photoconductive material on a ceramic substrate. The deposited material may be cadmium sulfide, cadmium selenide or cadmium sulfo-selenide. Metal contacts evaporated over the surface of the photoconductor serve as external electrical connections. These thin films of photoconductive material have a high sheet resistance.

145

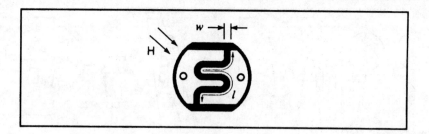

Figure 5-10. *An illustration of a photocell's electrodes which also control its resistance.*

The space between the two contacts (See *Figure 5-10*) is therefore narrow and inter-digitated (resembles interlocking fingers) for low cell resistance at moderate light levels. Note the width "W" between conduction pads and "l" which is the lineal length of the trace where the gap and conductor meet. There are also slightly different surface patterns. (See *Figure 5-11*) When using photocells you must ask yourself two questions. First, what kind of performance do you want, and secondly, what is the operating environment?

Sensitivity

This is a photodetector's relationship with the light falling upon it and the resulting output. This output is a controlled resistance. First, characterize the light source's intensity and spectral content. There are many curves for photocells on data sheets which do this.

Illumination is in both *fc* (footcandles) and *lux*. The light source standard is a tungsten incandescent lamp. This lamp's spectral composition of generated light matches that of a black body color temperature of 2,850 K.

Blackbodies

These are metal spheres which are theoretically perfect receivers and emitters of light or radiant energy. The 2850 K stands for 2,850 Kelvin, no degree sign used. 2850 - 273.15° C above absolute zero or a tungsten light's equivalent color, which equals heating a black body (2850 - 273.15) degrees Celsius above absolute zero.

Photocell manufacturers have made modifications to the chemical composition of the detector's materials. For a given photoconductor material, at a certain illumination level, the photoconductive material has a certain sheet resistance. The elec-

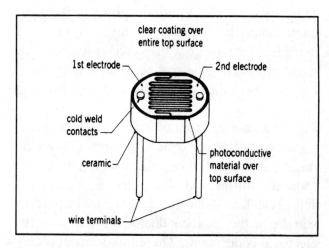

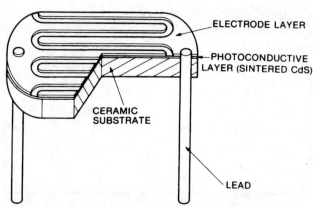

Figure 5-11. *A photocell's external physical structure.*

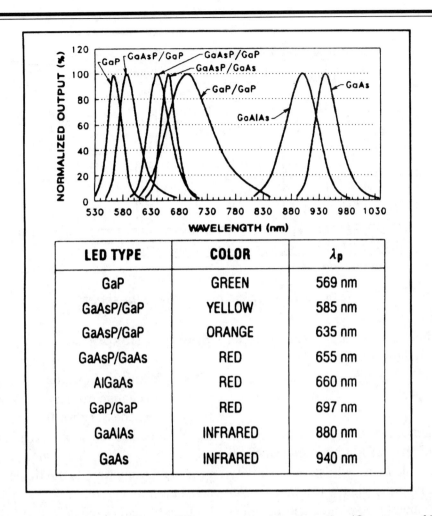

LED TYPE	COLOR	λ_p
GaP	GREEN	569 nm
GaAsP/GaP	YELLOW	585 nm
GaAsP/GaP	ORANGE	635 nm
GaAsP/GaAs	RED	655 nm
AlGaAs	RED	660 nm
GaP/GaP	RED	697 nm
GaAlAs	INFRARED	880 nm
GaAs	INFRARED	940 nm

Figure 5-12. *The output of an LED source vs. wavelength. (Courtesy of Vactec)*

trodes determine geometry. (See *Figure 5-10* again) You can vary the ratio of w/l over a wide range to achieve your design goals. Typical values for w/l range from 0.002 to 0.5, providing maximum flexibility for terminal resistance and maximum cell voltage.

Spectral Response

The relative sensitivity of a photocell is like the human eye in that it is dependent upon the wavelength (color) of the incident light. If you desire the spectral response of green, the peak emissions are about 560 *nm* and a type 0 photocell material with an exactly matched spectral response of 560 nm would be ideal. Type 1 photocell material has a 730 nm response and would not work here but readily detects red light. (See *Figure 5-12*)

Slope Characteristics

Plots of the resistance for photocells versus light intensity result in a series of curves with different slopes. Many applications require not only the absolute value of resistance at a given light level but also a set value of the resistance as the light source varies. You can specify this relationship with the parameter γ (small gamma), defined as a straight line passing through specific points on the resistance curve. (See *Figure 5-9* again) Vactec, a leading photocell manufacturer, uses 10 lux (0.93 fc) and 100 lux (93 fc) as their points.

In detecting a flame, your presence/absence criterion corresponds to an ON/OFF or digital output; therefore, a steeper slope is better. A photocell used

ELECTRO-OPTICAL CHARACTERISTICS @ 25° C (16 HRS. LIGHT ADAPT, MIN.) [4]

Part Number •	Old Part Number	Resistance (Ohms) [3]								Peak Spectral Response		Sensitivity (γ, typ.) LOG (R10:R100) LOG (100:10)	VMax (V, pk)	Response Time @ 1 fc (ms, typ.)	
		2 fc 2850 K			10 lux 2850 K			DARK		(nm)	Type			Rise (1-1/e)	Fall (1/e)
		Min.	Typ.	Max.	Min.	Typ.	Max.	Min.	sec.						
VT83A334	VT835	220k	330k	440k	—	700k	—	100M	5	550	3	1.0	300	35	5
VT83F603	VT83D	20k	60k	100k	—	120k	—	10M	5	550	3	0.95	200	35	5
VT83A333	VT833	22k	33k	44k	—	75k	—	15M	5	550	3	0.95	300	35	5
VT83E303	—	—	—	30k	—	25k	63k	100k	5M	10	550	3	0.95	100	35
VT83D223	—	11k	22k	33k	—	40k	—	5M	10	550	3	0.9	200	35	5
VT83C173	—	—	17k	—	20k	35k	50k	20M	10	550	3	0.95	200	35	5
VT83F902	VT83C	3k	9k	15k	—	18k	—	1M	5	550	3	0.95	200	35	5
VT83F302	VT83B	1k	3k	5k	—	6k	—	100k	5	550	3	0.95	100	35	5
VT84A901	VT841L	0.6k	0.9k	1.2k	—	1.8k	—	500k	5	675	4	0.85	100	35	18
VT87A302	VT871	2k	3k	4k	—	6k	—	10M	5	615	7	0.8	100	70	15

Table 5-1. *A typical photocell's data sheet.*

in an analog fashion, such as a light intensity meter, requires a much shallower slope.

Resistance Tolerance

You define the sensitivity of a photocell's resistance at a specific level of illumination. Since no two photocells are identical, you state sensitivity as a specific resistance, plus a tolerance. You specify both the value of resistance and its tolerance for only one light level. For moderate excursions from this specified light level, the tolerance limits remain almost constant. However, when the light level is decades larger or smaller than the reference level, the tolerance can greatly differ. As light levels decrease, the spread in their resistance tolerances increases. For increasing light levels, the resistance tolerance tightens.

Dark Resistance

This is naturally the resistance in total darkness. Some applications closely relate this parameter to leakage current at a specific voltage across the photocell. Too high a leakage can cause false triggering. You often specify the dark resistance as the minimum resistance expected 5 seconds after you remove the photocell from a light source of 2 fc intensity. Typical dark resistance values range from 500 kilohms to 200 megohms.

Speed of Response

This measures the speed at which a photocell responds to a change from light-to-dark. The rise time is the time necessary for the light conductance of the photocell to reach one time constant *tau* (τ). 1 - 1^e, or 63% of its final value (the same as a capacitor). Decay time is the time to reach one time constant *tau* (τ), 1/e, or 37%. At 1 fc of illumination the typical range is from 5 to 100 msec. (See the two extreme right columns in *Table 5-1*.)

The response speed depends on light level, light history, and ambient temperature. The faster responses occur with greater light. However, if you've stored the photocell in the dark for a long time, it takes longer to respond to light. Photocells are also slower in the cold.

A Photocell's Light History

All photoconductive cells have a phenomenon called hysteresis, light memory, or light history effect. Simply stated, a photocell tends to remember its most recent storage condition (light or darkness). Let's use an analogy of the human eye's sensitivity to light. This depends on what level of light and its duration your eyes recently experienced. You have experienced coming from the outdoors

MATERIAL TYPE	ILLUMINATION (fc)				
	.01	.1	1.0	10	100
TYPE 0	1.60	1.40	1.20	1.10	1.10
TYPE 1	5.50	3.10	1.50	1.10	1.05
TYPE 2	1.50	1.30	1.20	1.10	1.10
TYPE 3	1.50	1.30	1.20	1.10	1.10
TYPE 4	4.50	3.00	1.70	1.10	1.10
TYPE 7	1.87	1.50	1.25	1.15	1.08

Table 5-2. *Light history vs. resistance.*

on a bright summer day and being temporarily blinded under normal levels of illumination. Your eyes adjust after a certain amount of time elapses.

Table 5-2 shows the relationship between light history and light resistance at various light levels for different photocell materials. You determine these values by dividing the resistance of a given cell, following *infinite* light history (R_{LH}), by the resistance of the same cell following infinite dark history (R_{DH}). Realistically, 24 hours in the dark achieves R_{DH} and 24 hours in 30 *fc* of light achieves R_{LH}.

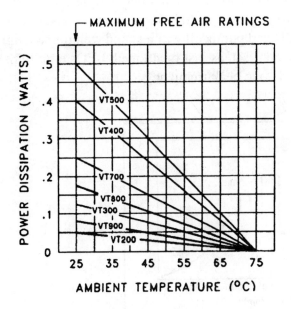

Figure 5-13. *A graph of a photocell's power dissipation vs. temperature. (Courtesy of Vactec)*

Maximum Power & Material Types

The maximum power any photocell can withstand is typically lower with lower voltage photocells. *Figure 5-13* shows power dissipation versus ambient temperature and demonstrates photocells derate linearly to zero at ambient temperature of 75°F. If you use a heatsink you can approximately quadruple photocell power ratings. *Table 5-3* shows the best material for your application. If you are an experimenter, you need to know how to measure light resistance and understand the spectral output of some common light sources.

Measuring Light Resistance

The industry standard source for light resistance is a tungsten filament lamp operating at 2850 K. Specifying the 2850 K color temperature defines its spectral output. Tungsten filament light has fixed amounts of blue, green, red and infrared light.

For consistency and ease of comparing different photocells, photocell manufacturers, such as Vactec, list light resistance values for photocells at two different standard light levels: 2 *fc* and 10 lux. The fc is the old historical English unit symbol for measuring light. The lux (the newer metric based unit) is the same, but the distance is one meter, instead of a foot. These equations convert between fc and lux:

1.0 fc = 10.76 lux
1.0 lux = 0.093 fc

149

MATERIAL CHARACTERISTICS (General Trends)				
CADMIUM SULFIDE		◄—————————————►		CADMIUM SELENIDE
TYPE 2 & 3	TYPE 0	TYPE 7	TYPE 4	TYPE 1
Lower	◄——————	**Temperature Coefficient**	—————►	Higher
Higher	◄——————	**Sheet Resistivity**	—————►	Lower
515 nm	◄——————	**Peak Spectral Response**	—————►	715 nm
Slower	◄——————	**Speed of Response**	—————►	Faster
Lower	◄——————	**Resistance Slope**	—————►	Higher
Smaller	◄——————	**Light History Effect**	—————►	Larger

Table 5-3. *A photocell material characteristic selection guide.*

Spectral Output of Common Light Sources

Incandescent lamps are black body radiators whose spectral output is dependent solely upon their color temperature. The sun has approximately the same spectral radiation distribution as that of a black body at 5900 K. However, the sun's spectrum contains H_2O and CO_2 absorption bands. (See *Figure 5-7* again)

Fluorescent and mercury vapor lamps exhibit a broad band spectral output with multiple narrow spikes in their spectra, unlike tungsten's quite predictable well-behaved spectral output. *Figure 5-14* plots the light output of a typical daylight type fluorescent tube. In contrast, forward biased LEDs, the light source of choice in most applications, emit a very narrow spectral bandwidth (light of one color). The "color" of emitted light depends on the LED semiconductor material. (See *Figure 5-12* again)

Spectral Matching of LEDs and Photocells

Since light sources and light detectors are almost always used together, they must transfer maximum energy at the same frequency to be the most efficient. If your design involves a light source with a broad band spectral output, you are assured your selected photocell will respond well to most light. In contrast, as just mentioned, LEDs emit light at only one peak frequency.

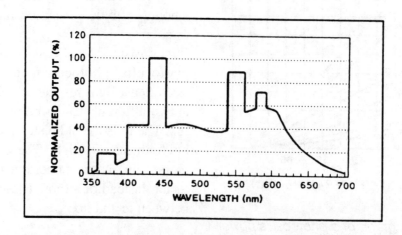

Figure 5-14. *The output of a fluorescent lamp vs. wavelength. (Courtesy of Vactec)*

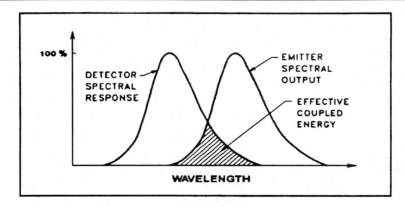

Figure 5-15. *The coupled energy of an LED emitter and a photocell detector. (Courtesy of Vactec)*

Figure 5-15 shows spectral matching factors of multiplying the optical detector's response curve by the LED's spectral output curve and then measuring the resulting coinciding area. The LED/photocell matching factors listed in *Table 5-4* are independent of power output from the LEDs. To approximate how well any LED/photocell combination works, consider the LED's power output at your circuit's forward current.

Millicandela is a photometric term for measuring LED intensity which assumes the human eye is the detector. For most detectors, other than the human eye, the most convenient system for measurement is the radiometric system. LEDs of a given type can show as much as a 5:1 manufacturing spread in power outputs. *Table 5-5* shows how you can factor in the electrical output power of the LEDs (10 mA), and derive a coupling factor.

Spectral Response Matching Applications

Let's match the spectral responses of an LED with a photocell for maximum power transfer and then follow it with two practical circuits:

1. A lamp controller.
2. A film speed monitor.

LED TYPE	λ_p (nm)	TYPE 0	TYPE 1	TYPE 2	TYPE 3	TYPE 4	TYPE 7
GaP	569	39%	3%	34%	40%	16%	21%
GaAsP/GaP	585	60%	6%	39%	52%	28%	46%
GaAsp/GaP	635	49%	12%	27%	38%	43%	54%
GaAsP/GaAs	655	31%	7%	15%	27%	41%	36%
AlGaAs	660	31%	8%	16%	23%	30%	35%
GaP/GaP	697	47%	61%	23%	31%	100%	67%
GaAlAs	880	—	13%	—	—	10%	1%
GaAs	940	—	4%	—	—	3%	—

Table 5-4. *Normalized LED/photocell matching factors.*

LED TYPE	COLOR	λ_p	POWER OUTPUT	
			I_f = 1 mA	I_f = 10 mA
GaP	GREEN	569 nm	1.2 μW	24.1 μW
GaAsP/GaP	YELLOW	585 nm	.3 μW	26.2 μW
GaAsP/GaP	ORANGE	635 nm	3.2 μW	101.9 μW
GaAsP/GaAs	RED	655 nm	6.2 μW	102.1 μW
AlGaAs	RED	660 nm	33.8 μW	445.1 μW
GaP/GaP	RED	697 nm	54.3 μW	296.2 μW
GaAlAs	INFRARED	880 nm	76.8 μW	1512.3 μW
GaAs	INFRARED	940 nm	35.5 μW	675.0 μW

Table 5-5. *Normalized LED coupling factors at 10 mA.*

The Lamp Controller

Often you have to control the light output from an incandescent light. You can easily monitor the lamp's output using feedback circuitry. (See *Figure 5-16*) Next, select the proper photocell, realizing the design is a simple voltage divider biased so the logic inverter's threshold tripping point is beyond the upper and lower logic limits. Select the proper photocell on the basis of temperature, power and spectral response.

An incandescent lamp's spectral output is very wide (covering both the visible and infrared spectrum), but a photocell's response is not that critical. Since it should be low cost you'll select a plastic coated photocell. And since it operates outside from -20° C to 60° C you'll want to compare the relative resistance versus temperature. Power dissipation, P_D (max) for the photocell when R_{ON} is 1 kΩ, P_D ((max) = 6 volts) • (6 volts/1kΩ) = 36 mW. Comparing device packages (See *Table 5-6*), let's use the VT82A192 photocell.

Film Speed Monitor

This application detects a 0.20" square in a moving strip of a film negative traveling at 4"/sec. An LED at the other side of the strip illuminates the photocell. (See *Figure 5-17*) When the R_{ON} resistance of the photocell drops below 12 kΩ, the output of the comparator goes high. This signals presence of the square hole. You can adjust the forward current drive to the LED so the R_{ON} of the photocell reads 8 kΩ with no film present. There are two photocells you might chose. (See *Figure 5-18*)

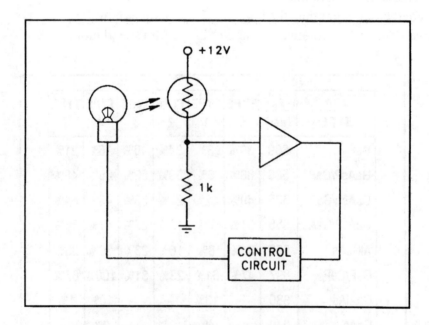

Figure 5-16. *A photocell lamp controller circuit.*
(Courtesy of Vactec)

Package	P_D @ 25° C	Derate above 25° C	P_D @ 60° C
VT900	80 mW	1.6 mW / ° C	24 mW
VT800	175 mW	3.5 mW / ° C	52.5 mW

Table 5-6. *Commercially available photocell parameters.*

The following equation calculates the illumination time when you know the linear velocity of the film and the size of the hole for the photocell:

$$T_{LIGHT} = (0.20") \cdot (1 \text{ sec}/4") = 50 \text{ msec.}$$

The R_{ON} of the photocell must drop below 12 kΩ within 50 msec to detect the hole. Type 0 photocell is inappropriate because after 0.078 secs., its R_{ON} is:

$$R_{ON} (0.078 \text{ sec.}) = (1/.63) \cdot (8 \text{ k}\Omega) = 12.7 \text{ k}\Omega$$

This photocell is much too slow to trigger the op amp comparator circuit. But the R_{ON} resistance of the VT21A782 photocell drops to 12.7 kΩ within 0.015 sec. and is plenty fast for this application.

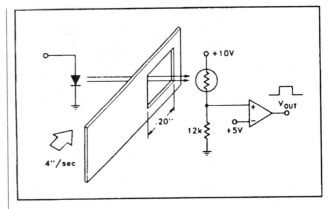

Figure 5-17. *A photocell film speed monitor. (Courtesy of Vactec)*

PHOTODIODES

Silicon Photodiode Structure

Silicon photodiodes are rugged and stable over time. You may also use multiple layer glass filters to correct them to the CIE photopic curve for photometric application. For radiometric applications you may use filters to obtain a flat response from 450 to 950 nm. Silicon photodiodes are made from a single crystal silicon wafer similar to those in ICs. Photodiodes however require much higher purity silicon. The purity of silicon is directly re-

Performance		Full Scale	Gain (V/A)	Bandwidth
		2mA	10^3	30KHz
		200µA	10^4	30KHz
		20µA	10^5	30KHz
		2µA	10^6	30KHz
		200nA	10^7	30KHz
		20nA	10^8	10KHz
		2nA	10^9	500Hz
		200pA	10^{10}	100Hz
Output	± 2V, Full Scale			
Accuracy	± 1% Full Scale, All Ranges			
Noise	0.1% rms Full Scale			
Zero Adjust	0 to -2V, Display Only			
Calibration Adjust	1X to 10X, Display Only			

Table 5-7. *The ranges and gain accuracies of a typical, commercially-available photodiode amplifier*

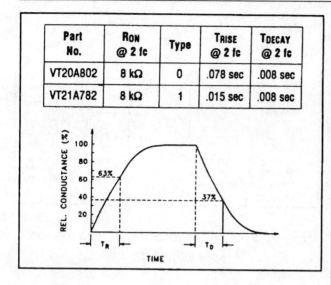

Part No.	R$_{ON}$ @ 2 fc	Type	T$_{RISE}$ @ 2 fc	T$_{DECAY}$ @ 2 fc
VT20A802	8 kΩ	0	.078 sec	.008 sec
VT21A782	8 kΩ	1	.015 sec	.008 sec

Figure 5-18. *Two photocells and a relative conductance graph. (Courtesy of Vactec)*

Small metal contacts are applied to the front surface of the device and the entire back is coated with a contact metal.

The back contact is the cathode, the front contact is the anode. The active surface is coated with either silicon nitride, silicon monoxide or silicon dioxide for protection and anti-reflections. You optimize this coating's thickness for wavelength dependent applications.

The characteristics of P-N junctions are well known. However, photodiode junctions are unusual because the top "P" layer is very thin. Near the P-N junction the silicon becomes depleted of electrical charges. You can vary the depth of this "depletion region" by applying a reverse bias voltage across the junction. When the depletion region reaches the back of the diode, the photodiode is fully depleted. The depletion region is important to photodiode performance since most of the sensitivity to radiation originates there.

lated to its resistance, with higher resistance indicating greater purity. Centronic, a manufacturer of silicon photodiodes, uses silicon with a resistance in the range of 10 Ω/cm to 10,000 Ω/cm.

The Photodiode's Structure

Figure 5-19 is a cross section of a typical photodiode. The process starts with N-type material and a thin P-type layer is formed on the surface of the device by thermal diffusion or ion implantation of the appropriate doping material (usually boron).

The capacitance of the P-N junction depends on the thickness of this variable depleted region. Increasing the bias voltage increases the depth of this region and lowers capacitance until reaching the fully depleted condition. Junction capacitance is also a function of the resistivity of the silicon used and the active area size of the photodiode. (See

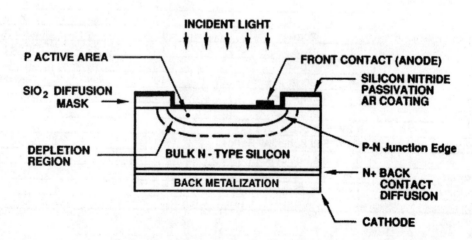

Figure 5-19. *The internal structure of a photodiode. (Courtesy of Centronics)*

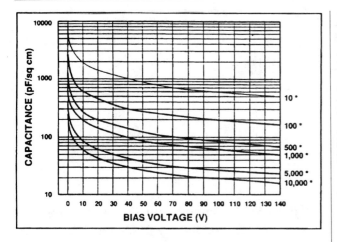

Figure 5-20. *Silicon photodiode capacitance vs bias voltage. (Courtesy of Centronics)*

Figure 5-20) Note also in *Figure 5-21* how greatly the area on the surface of various photodiodes differs.

When the active area absorbs light, it forms electron-hole pairs. The electrons and holes are separated with the electrons going to the "N" region and the holes going to the "P" region. This causes a light generated current (usually abbreviated as I_{SC}). This migration of electrons and holes to their respective places is the *photovoltaic effect*.

Silicon photodiodes are the most useful as current generators, although illumination also generates a voltage. You need to convert the photodiode cur-

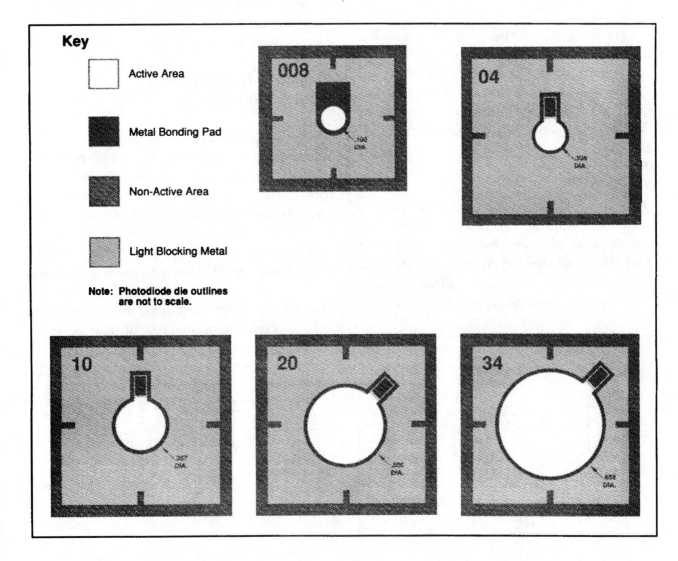

Figure 5-21. *Various shapes and sizes of silicon photodiode dies. (Courtesy of Centronics)*

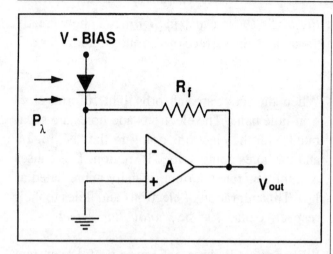

Figure 5-22. *A negatively biased photodiode amplifier. (Courtesy of Centronics)*

rent produced into a voltage with a photodiode amplifier. This current-to-voltage conversion allows you to use (for example) an oscilloscope (a voltage sensing device) to view the signal. *Table 5-7* shows typical parameters for a commercially available photodiode amplifier from Centronic.

Biasing

When you apply reverse bias, some current flows without illumination. This is "dark current". Photodiodes have very low resistance when you forward bias the anode and very high resistance when you reverse bias the anode. (See *Figures 5-22* and *5-23*.) In the photovoltaic and zero bias modes

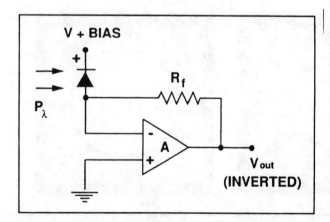

Figure 5-23. *A positively biased photodiode amplifier. (Courtesy of Centronics)*

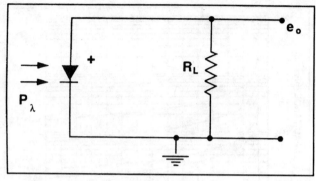

Figure 5-24. *A photodiode in the photovoltaic mode. (Courtesy of Centronics)*

(See *Figures 5-24* and *5-25* respectively), a semiconductor photodiode generates current or voltage in the photodiode's forward direction. Therefore, the generated polarity is opposite to that required for the biased mode.

Biasing of Non-Semiconductor Vacuum Photodiodes

Figure 5-26 shows the Hamamatsu UVtron™ UV flame and fire detecting non-semiconductor vacuum photodiode biased correctly and incorrectly. Note the normal and highly "smashed down" resulting angular sensitivity patterns in *Figure 5-27*. The anode should be high and the cathode less positive or grounded. (See *Figure 5-28*) The black dot within the photodiode indicates it is gas filled.

The UVtron™ is also well suited for detecting high voltage discharge phenomena such as the invis-

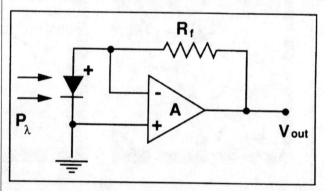

Figure 5-25. *A photodiode in the zero bias mode. (Courtesy of Centronics)*

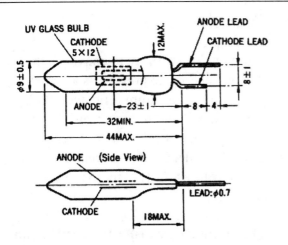

Figure 5-26. *The Hamamatsu's UVtron^tm UV photodiode. (Courtesy of Hamamatsu)*

ible corona discharge of high voltage transmission lines. The UVtron^tm detector can detect a cigarette lighter in a room at greater than 15 feet. The UVtron^tm ultraviolet detector uses its metal's photoelectric effect and the gas multiplication effect to produce a narrow sensitivity band of 185 to 260 nm. This is the highly energetic and potentially most biologically damaging UV-C region of UV light. (See *Figure 5-29*) In nature, the oxygen in the stratosphere totally absorbs UV-C and it fortunately never reaches the Earth's atmosphere.

This UV photodiode is totally insensitive to visible light and therefore requires no optical filters. Since the UVtron^tm photodiodes also emit UV, don't place them too close to each other or they will optically interference with each other.

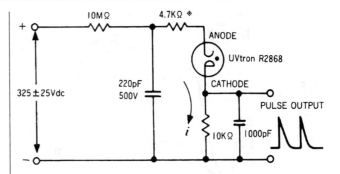

Figure 5-28. *Properly biasing the UVtron^tm. (Courtesy of Hamamatsu)*

Responsivity

Measuring the sensitivity of a photodiode established a ratio of radiant energy (in watts) incident on the photodiode to the photocurrent, expressed in Amps. Radiant energy has the unit symbol of watts/cm^2 and photodiode current uses Amps/cm^2. The terms cm^2 cancel, simplifying the ratio to just amps/watts (A/W). Do not confuse this with spectral and angular sensitivities. These describe:

1. The wavelength of the light to which the photodiode best responds.
2. The photodiode's angular span of directivity.

Spectral Response

Figure 5-30 illustrates silicon becomes transparent to radiation no longer than 1,100 nm. The manufacturer usually applies some surface coat-

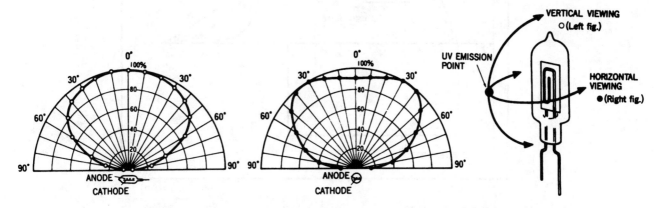

Figure 5-27. *The Hamamatsu's UVtron^tm Angular sensitivity patterns from proper and improper bias. (Courtesy of Hamamatsu)*

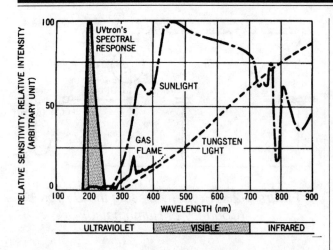

Figure 5-29. *The UVtron's highly UV spectral response. (Courtesy of Hamamatsu)*

ing to enhance the spectral response and reduce the effects of a scratched surface, which is microscopic in magnitude, but still inevitable. Coatings also enhance efficiencies at the wavelength of choice but attenuate other wavelengths, forming, in effect, an optical filter. Standard glass absorbs wavelengths less than 300 nm; therefore, UV emitters and detectors use a fused silica or a special UV transmitting glass. (See *Figure 5-31*)

Quantum Efficiency

A photodiode's capability to convert light energy to electrical energy, expressed as a percentage, is the photodiode's quantum efficiency (Q.E.). The sensitivity of a photodiode may also be expressed in practical units of amps of photodiode current per watt of incident illumination. The following equation relates Q.E. to the photodiode's responsivity:

$$Q.E. \ (\%) = 1.24 \cdot 10^5 \cdot R \ (A/W)/\lambda \ (nm)$$

Operating under ideal conditions of reflectance, crystal structure, and internal resistance, a high quality silicon photodiode approaches 80% Q.E.

Temperature Effects

The operating characteristics of a photodiode results in two distinct changes in *operating temperature*. The first change is a shift in the Q.E due to changes in the radiation absorption coefficient of the photodiode. Q.E. values shift lower in the UV range and higher in the IR region. There is virtually no shift in the visible light range. (See *Figure*

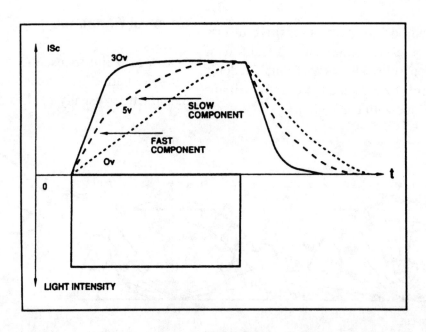

Figure 5-30. *Radiation penetration in silicon at various wavelengths.*

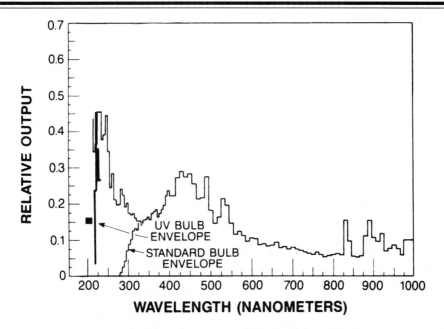

Figure 5-31. *Spectral output of a standard glass versus a UV bulb.*
(Courtesy of Hamamatsu)

5-32) The second change is the exponential increase in the thermally excited electron-hole pairs. This results in increased dark current. The leakage though doubles for each 8 to 10°C, just like in a normal semiconductor.

Noise Equivalent Power (NEP)

The NEP is expressed in the following equation:

NEP = noise current (A)/responsivity (A/W)

The NEP depends upon the bandwidth measuring system. To remove this dependence, divide the figure by the square root of the bandwidth. This yields NEP in watts/Hz$^{1/2}$. Since the photodiode light power to current conversion depends on wavelength radiation, its NEP corresponds to a particular wavelength. The NEP is nonlinear over the wavelength range, just like responsivity.

The noise generated by a silicon photodiode, operating under reverse bias, is a combination of shot noise, due to dark leakage current, and Johnson noise due to the shunt resistance of the device and

the ambient temperature. The shot noise current produced by the reverse leakage current is:

$$I_S = (2e\ I_D\ B)^{1/2}$$

I_S is shot noise, e = electronic charge (1.6 • 10^{-19} Coulombs), I_d = dark leakage current in amps and B = bandwidth of the system in Hz.

If a photodiode has I_D = 2 nA and a shunt resistance of 5 • 10^8 Ω, and a responsivity of (Rλ) = 0.5 A/W, which is typical of most photodiodes, and letting B = 1 Hz, we have:

$$I_S \text{ (shot noise)} = 2.5 • 10^{-14} \text{ A}$$
$$I_J \text{ (Johnson noise)} = 5.6 • 10^{-15} \text{ A}$$
$$I_N \text{ (total noise)} = 2.6 • 10^{-14} \text{ A}$$
$$\textbf{NEP} = 5.1 • 10^{-14} \textbf{ Watts}$$

Response Time

This is often the most important parameter of a photodiode and consists of two elements. The first is the root mean square sum of charge collection time and the RC time constant tau (τ) arising from

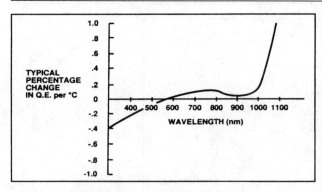

Figure 5-32. *The temperature dependence of Q.E.*

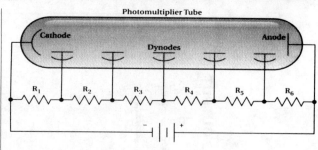

Figure 5-34. *Ensuring a set voltage distribution by using a voltage divider with a photomultiplier tube. (Courtesy of Keithley)*

the series plus load resistances and the junction and stray capacitances.

Charge collection time is voltage dependent and composed of a fast and a slow component. The fast component is the transit time of charge carriers (electrons and holes) through the depletion zone, at their respective drift velocities, under the influence of an electric field.

The slow component comes from photon energy absorbed outside the depletion region, producing carriers that are collected by diffusion. The transit time of these carriers will be relatively slow. The following equation relates switching times to frequency response.

$$f_c = 0.35/t_r$$

f_c is in Hz and t_r is in seconds.

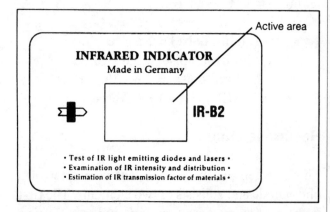

Figure 5-33. *An IR indicator card which emits visible light in proportion to the IR light intensity striking it. (Courtesy of Siemens)*

When a photodiode is operated in the unbiased mode, the slow diffusion component dominates, yielding risetimes on the order of 0.5 msec.

Photodiode Applications

The equivalent circuit of a photodiode has junction capacitance depending on the depletion layer depth and therefore the bias voltage. The value of the shunt resistance is usually high (in the $M\Omega$ region). The series resistance is low. The photocurrent generated to flow through R_D produces a voltage across the photodiode. The light is a logarithmic function of incident light intensity. The major disadvantage of this circuit is the signal depends on R_D, which typically has a wide spread of values over different production batches.

Overcoming Load Resistance

Feeding the photocurrent to an op amp's virtual ground is one way to achieve sufficiently low load resistance while having an amplified output voltage. This circuit has a linear response and has low noise due to the almost complete elimination of leakage current. In the photoconductive mode, the generated photocurrent produces a voltage across a load resistor, in which the shunt resistance is in parallel.

Since R_D is virtually constant in the reverse biased mode, use large values of R_L which will still give a linear response between the output voltage and applied radiation intensity (light). This form of cir-

Figure 5-35. *A commercially available electrometer picoammeter.*
(Courtesy of Keithley)

cuit is required for high speed response photodiode applications. The main disadvantage is the increased leakage current due to the bias voltage, which produces higher noise than the other types of circuit modes we've described. Refer again to *Figure 5-22* for a practical photodiode mode circuit and note the reverse biasing.

IR Sensing Cards

One inexpensive and quick method of testing an IR photodiode or an IR laser's output is to use an IR indicator card, see Appendix B for two sources. You can use the Siemens product (See *Figure 5-*

33) to estimate both IR intensity and radiation patterns. The active light receiving area of the IR-B2 is composed of Europium and Samarium, rare earth elements. It responds best to a $\lambda = 480$ *nm.* which is blue light. However, under IR excitation, the card emits visible light, the brightness of which is directly proportional to the IR's intensity. There is an IR element within a multiple sensitivity peaking fluorescent light. Therefore, avoid light source environments containing IR or you will distort the IR indication of your intended to be measured IR light source. To guard against this, use light filters from Schott, part numbers RG780 and RG1000, which isolate specific segments of the IR spectrum.

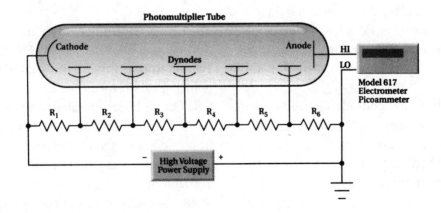

Figure 5-36. *A photomultiplier tube configured for a negative current reading.*
(Courtesy of Keithley)

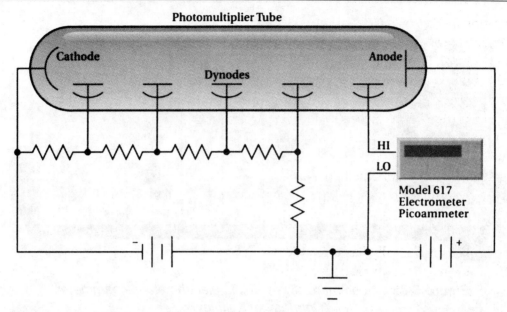

Figure 5-37. *Reconfiguring a photomultiplier tube for a positive current reading. (Courtesy of Keithley)*

These cards charge in a few seconds and discharge or have rise and fall times for emission in the range of microseconds. Since these cards have a self charging effect, it takes months in total darkness to completely discharge them, a lot longer than a photocell! Also, in time, their discharge or light emission diminishes and they eventually just wear out.

Photomultiplier Tubes

Unlike the far less sensitive photodiode, a photomultiplier tube converts light energy into electric current through both photoemission and secondary emission. Secondary emission causes surface electrons to be released. Photoemission is when electrons are released after photons strike the photocathode.

A photomultiplier tube consists of a light-sensitive cathode emitting electrons in proportion to the photons striking it. These electrons accelerate to the next stage where they cause emission of three to six secondary stages (called dynodes), depending on the tube type. Gains in the range of one million are common.

Electrons accelerate at each successive stage toward a curved plate (the dynode), which is more positive than the previous one. Using a voltage divider ensures this, and also equitable voltage distribution. (See *Figure 5-34*) The voltages applied to each dynode are a function of the photomultiplier tube design. The total resistance of the dynode resistors or "string" should be equal to the current flowing through the total series resistance and also be at least 100 times the expected anode current of the tube as follows:

$$\text{R}_\text{T} = \text{Voltage (anode-to-cathode)}/ (100 \cdot \text{anode current})$$

Most photomultiplier tubes require an anode-to-cathode potential in the 1,000 to 3,000 volt range. The photomultiplier tube's anode is its readout point and therefore operates near ground potential. The cathode operates at a high potential. The anode current of most photomultiplier tubes range from picoamperes up to 1 mA. When measuring this small current you will commonly use an electrometer picoammeter. (See *Figure 5-35*) The low input voltage drop of such an instrument keeps the anode at virtual ground potential.

The configuration in *Figure 5-36* reads a negative current; however, there may be times you wish to read a positive current. By rearranging the way the photomultiplier tube connects to the test instrument and by introducing a power source, you can have a positive current reading. The picoammeter (See *Figure 5-37*) reads the current at the last dynode, which is equal to the anode current minus the current flowing to the last dynode. You therefore sacrifice a slight amount of gain to obtain this positive reading.

The photomultiplier tube usually has a small current flowing, even when the cathode does not illuminate. This is the tube's dark current and is not significant in most measurements. In other cases you can subtract this from your final reading. Also, there are test instruments which allow you to cancel out this zero suppression feature.

Chapter 5 Quiz

1. Which can switch faster?
   A. A photocell.
   B. A photodiode.
2. A photocell's resistance versus light can vary as much as:
   A. 10:1.
   B. 20:1.
   C. 50:1.
   D. Several orders of magnitude.
3. Bulk effect photoconductors have:
   A. No leads.
   B. No junctions.
   C. No light memory.
   D. No resistance.
4. The relationship between incident light intensity and the photoconductor's output is:
   A. Its reverse transfer function.
   B. Its resistance.
   C. Its sensitivity.
   D. Its power rating.
5. A photoconductor's slope characteristic is its resistance to its light intensity, T or F.
6. If you _____ a photoconductor's light you improve its resistance tolerance.
   A. Increase.
   B. Decrease.
7. Dark resistance is a _____ resistance in the dark.
   A. Photocell's
   B. Photodiode's
8. _____ describes a photocell's tendency to remember the lighting intensity of its most recent storage environment.
   A. Light memory
   B. Light history
   C. Hysteresis
   D. All of the above.
9. Using a Tungsten standard light to measure a photocell's light resistance poorly defines the spectral output, T or F.
10. Spectral matching of a photocell to an LED ensures maximum transfer of energy, T or F.
11. _____ require even purer silicon than an ordinary semiconductor.
    A. Photocells.
    B. Photodiodes.
12. Applying a variable reverse bias voltage across a photodiode's junctions controls its depth of depletion, T or F.

13. Photodiodes are most useful as current generators but require:
    A. A series dropping resistor.
    B. A current-to-voltage converter.
    C. A photodiode amplifier.
    D. B and C.

14. The coatings on photodiodes attenuate wavelengths other than those of its purposeful design, and form, in effect _____.
    A. A micro-mirror.
    B. An optical filter.
    C. An optical amplifier.
    D. None of the above.

15. Reverse biasing a photodiode generates noise composed of thermal or Johnson noise and _____.
    A. Shot noise.
    B. Audible noise.
    C. Cryogenic noise.
    D. None of the above.

16. A photomultiplier tube uses:
    A. Photoemission only.
    B. Secondary emission only.
    C. Both A and B.
    D. None of the above.

17. UV-C:
    A. Occurs naturally within our atmosphere.
    B. Is not very energetic.
    C. Is not potentially hazardous.
    D. None of the above.

18. The black dot within UVtron^{tm}'s vacuum photodiode indicates:
    A. It is gas filled.
    B. The direction of sensitivity.
    C. The maximum voltage is 100 volts even.
    D. Just A and B.

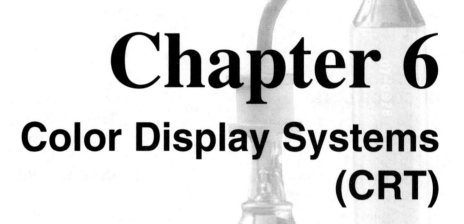

Chapter 6
Color Display Systems (CRT)

Chapter 6
Color Display Systems (CRT)

There is nothing which gives more life to certain images, and even text, than a color display. In a color display system, the CRT predominately determines visual characteristics — more so than all other factors combined!

THE THREE MAIN TYPES OF COLOR CRTS

The types of color CRT tubes are:

1. The Penetron Tube.
2. The Color Write Through Tube.
3. The Shadow-Mask Tube.

The Penetron Tube

This resembles a monochrome CRT; however, it has two phosphors. These are distributed in separate layers. (See *Figure 6-1*) The two phosphors are red and green. Each layer of phosphor requires a different energy level to activate it. In one method, the red phosphor acts as an energy barrier which the beam must *penetrate* to reach the green phosphor, thus its name. Although it is theoretically possible for such a tube to have more than two layers of colored phosphors, it has proven impractical to implement. A low energy electron beam (6 kV) excites only the outer red phosphor and the dead layer stops it. A high energy electron beam (12 kV) penetrates the dead layer and excites the green phosphor.

Obtaining Different Penetron Beam Energies

You can do this in a number of ways; however, two ways are in common use. First, you can switch the target potential. In the second method, you can use two electron guns at different potentials. You simply turn current on in one or the other gun. In either system, each color requires a different deflection voltage or current to deflect the beam to a particular point on the screen.

The Color Write Through System

This second type color CRT is a direct view storage tube (DVST). *Figure 6-2* shows this tube has a writing gun operating at a large negative potential (6 kV) with respect to the target. The tube also has an array of low voltage flood guns and a special phosphor target.

An insulating layer, pierced with an array of conductive dots, separates the phosphor from a transparent conductor. Like the Penetron tube, there are two phosphors present. It mixes small particles of each phosphor. The normally green storage phos-

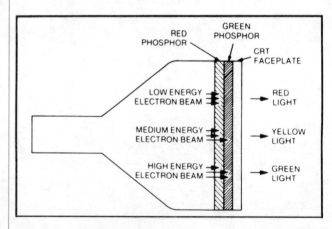

Figure 6-1. *The Penetron phosphor color CRT.*

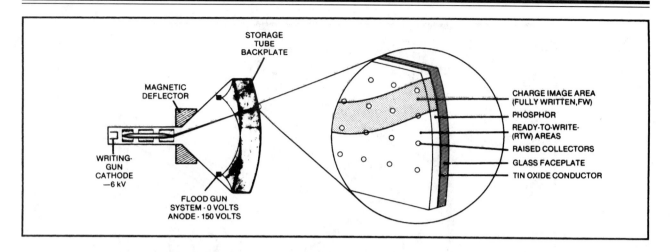

Figure 6-2. *The color write through DVST (Direct View Storage Tube).*

phor mixes with red phosphor particles, surrounded by a dead layer. The DVST's normal storage operation is unaltered by this phosphor arrangement.

The flood guns, which continuously flood the entire phosphor target with electrons, are at ground potential. They stay near ground potential through secondary emission. High energy beams produce a secondary emission greater than unity, the writ-

ten to areas charge to a potential of about +300 volts with respect to the flood-gun cathode. In the DVST's storage mode, the writing gun scans the target and writes by leaving charges on the phosphor dielectric. When the flood electrons strike the written target area, they cause the phosphor to luminesce. The unwritten areas, at or near ground potential, do not luminesce.

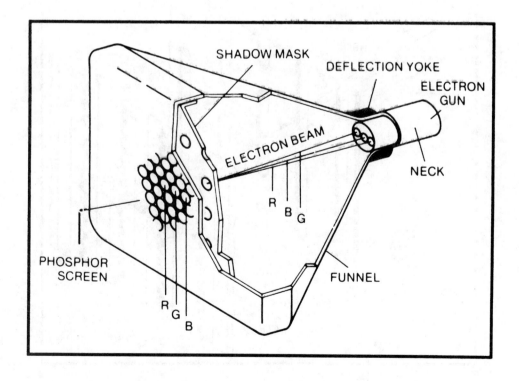

Figure 6-3. *The shadow mask color CRT.*

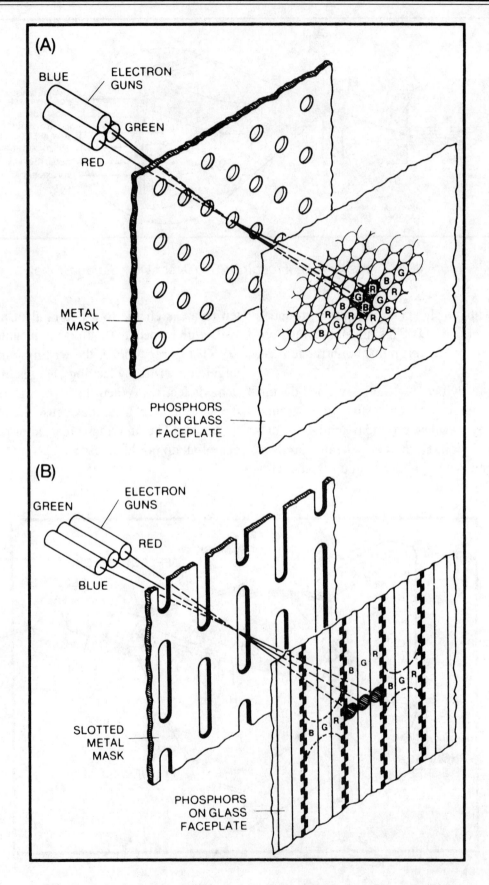

Figure 6-4. *A delta gun (A) and an in-line triad arrangement (B).*

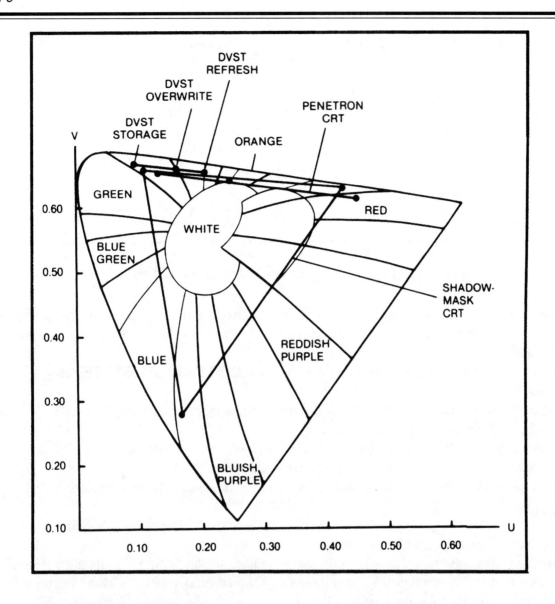

Figure 6-5. *The color gamut of present CRT color technologies.*

In preparation of writing to the screen, the written to areas are erased by pulsing the conductive back plane. This resets the phosphor potential to its lower bistable state. Decreasing the writing beam current below a certain threshold disables information storage. This coincides with phosphors briefly illuminating from the writing gun's high energy. This phenomenon is "write through." With the penetration phosphor in place, the writing beam's high energy electrons penetrate the red phosphor's dead layer and excite both phosphors. This produces a yellowish green trace. This is "color write through," thus the name of this color CRT.

While this occurs, the flood electrons are at a much lower energy state and excite only the green phosphor where the image is stored. Only one color is available for stored images but you can obtain other colors in the non-storage mode or by writing over the stored trace with the write beam.

The Shadow-Mask Tube

This the most popular, and unfortunately the most complex color CRT (See *Figure 6-3*) in principle of operation. The color mask tube's three electron guns address either three primary color phosphor

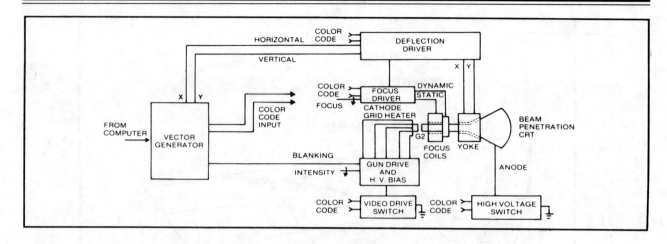

Figure 6-6. *A vector refresh display system.*

dots or three primary color stripes. The dots and stripes groupings form triads. These closely packed triads appear as a single dot. Color results from proportionally mixing the luminescence from the triad's individual dots or stripes. The shadow mask ensures each beam addresses only its assigned color dot or stripe. The beams from the red, green, and blue guns must be properly angled to pass through the shadow mask openings and strike the corresponding phosphor while all other phosphor dots are shadowed.

The dot pattern permits smaller horizontal spacing between triads; therefore, dots provide maximum resolution. Guns are typically in a delta or triangle configuration for dots and in-line for stripes. (See *Figure 6-4a* and *6-4b*) It is possible to use in-line guns with dot phosphors, but a delta arrangement requires less convergence circuitry.

Misconvergence mostly occurs when the three beams pass through the deflection yoke at slightly different angles and locations. This deflects them to slightly different spots on the screen. Correction circuitry though ensures the three primary colors' proper alignment or registration. You can avoid misalignment by displaying all three colors one at a time as Tektronix did with their DAS color systems. Another unique feature was the red, green and yellow CRT dots. The Penetron CRT and DVST with CWT are limited to a maximum of

three distinguishable colors. Only the shadow mask CRT has a full range of colors. (See *Figure 6-5*)

COLOR DISPLAY SYSTEMS

Three popular color graphics CRT systems are:

1. The vector storage.
2. The refresh vector.
3. The raster refresh systems.

The vector storage system requires a DVST. Theoretically, both the refresh-vector and the refresh-raster displays could use either the Penetron or the shadow-mask type CRT. However, the refresh vector is simpler with the Penetron type.

Refresh Vectors

Figure 6-6 illustrates the principle of operation. Deflecting the beam between the specified end points draws the vector. Images from a combination of vectors rapidly change by merely changing the vector's end points. Great dynamic capabilities result from relatively few points defining images consisting mainly of lines. However, you must refresh (repeat) the complete image to perceive a constant luminance deflection speed. This usually limits the number of vectors you can draw before flicker becomes apparent.

Penetron CRTs require a different deflection amplifier gain for each of the two colors; therefore, you often use field-sequential operation. This is writing red information in the first field, then the deflection amplifier gain changes before green information writes to the second field. You obtain a third color by overlapping (registering); however, this is difficult, sometimes requiring a third acceleration potential in a third field.

Vector Storage

Storing vectors on the screen storage display overcomes the flicker imposed limit on the number of vectors, something inherent in a refresh-vector system! *Figure 6-7* shows a typical vector-storage display system using a DVST with color write-through (CWT). Note its similarity to the refresh-vector system; however, deflection speed is not critical when the DVST is in the storage mode because of screen stored green vectors. This allows deflection speed to now only affect the time required to draw a complete graphics image. Drawing a limitless number of vectors causes no flicker. That's why the DVST is an excellent choice in drawing complex intricate images.

Combining the capabilities of both the DVST and CWT allows adding images with another color to the display. Operating the write beam at reduced currents prevents storage and a yellowish-orange spot appears on the screen. You can reflect this non-stored spot to produce refreshed vectors. Maximum deflection speed and flicker limits the number of vectors in this second color.

Writing the refresh vector on top of an identical stored vector produces a third greenish-yellow color. Unlike the Penetron, no misregistrations occur since the same writing beam potential is used in all operating modes. While a stored image must be erased all at once, a refreshed image can be selectively updated and even provide dynamics, such as in displaying the motions of a mechanism in action.

Raster Refresh

The refresh vector is the most common color graphics system. *Figure 6-8a* is an example of this system, with a shadow-mask CRT. This system uses three beams deflected together over the phosphor screen in a predetermined raster pattern. (See *Fig-*

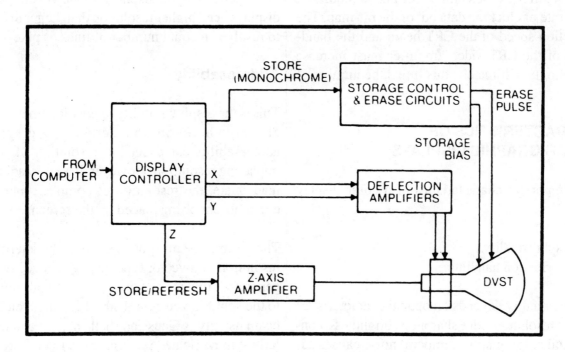

Figure 6-7. A vector storage display system.

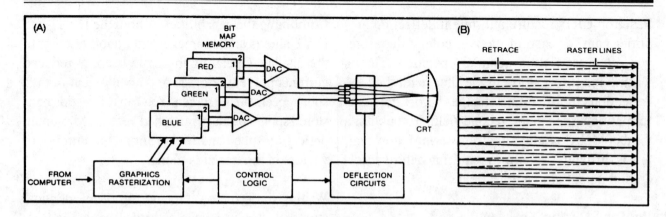

Figure 6-8. *A raster refresh display system (A) and a raster scanning pattern (B).*

ure 6-8b) A bit-map memory determines the time and current magnitude each of the three guns receives. This allows control of how much of each color is produced at each pixel or addressable point on the screen.

The information in the bit-map must be read out repeatedly at a rate fast enough to avoid flicker. How fast scan conversion can reload a bit-map determines the time required to change images on the screen. Larger bit-maps naturally slow this process. Therefore, raster scan images with many pixels must trade off speed of interaction with dynamic images. An increased number of pixels requires a faster rate of clocking data out of the bit-map. The deflection speed of the CRT beam and the bandwidth of the CRT video amplifier must increase accordingly. Ultimately, this limits the number of pixels.

CHARACTERISTICS OF COLOR GRAPHIC DISPLAYS

Two important characteristics of a color display are:

1. Display quality.
2. Information handling.

Display quality depends on optical characteristics such as resolution, edge sharpness, brightness, contrast and color quality. Temporal noise causes adverse effects such as flicker, jaggies and moire pat-

terns (to be explained shortly.) Information handling factors include display size, number of vectors or pixels, number of colors, and interactivity.

IMAGE QUALITY CHARACTERISTICS

Resolution affects image quality more than any other factor. Raster display discussions often incorrectly use resolution as being synonymous with the number of scan lines (addressability). Resolution is the display's ability to resolve — that is, to separate two closely spaced points or lines. Resolution primarily determines image sharpness. It is not dependent upon display size; however, smaller displays need higher resolution than larger displays to resolve an equal number of lines or pixels.

Addressability

This is the display's ability to position lines or pixels anywhere on the screen. A display's addressability can exceed its resolution while not impacting resolution. However, if addressability is inadequate, the resolution of complex images is either absent or misplaced on the screen.

The electron beam spot size primarily determines resolution of a vector type display. As vectors are generated, images consist of lines equal in width to the spot. The current distribution in an electron beam usually is Gaussian (bell shaped) and circularly symmetrical. (See *Figure 6-9*) There are several methods to specify display resolution; how-

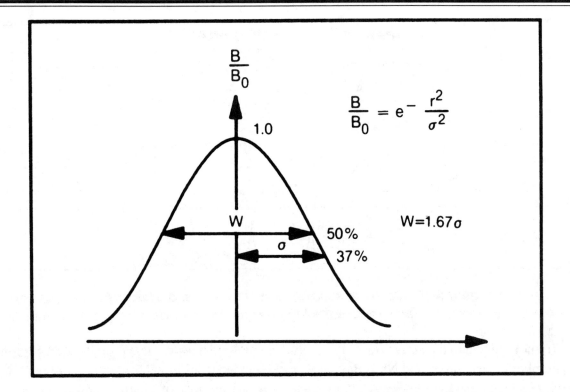

Figure 6-9. *A Gaussian spot's profile.*

ever, the comprehensive modulation transfer frequency (MTF) method considers not only spot size but also the minimum spacing between spots. Think of images as composed of a set of pairs of different spacings (spatial frequencies) and contrast. They are very similar to all electrical signals — a set of different amplitude and frequency sine waves.

The MTF measures how well a display passes the different spatial frequencies in an image. MTF plots contrast versus spatial frequency. Spatial frequencies with large MTFs indicate good resolution.

Displaying a raster of lines and then shrinking the raster until you can no longer resolve the lines is the most common CRT resolution measuring method. You can then divide the raster height by the number of lines to determine how closely you can place the lines and still provide an acceptable image. This method only yields the upper limit of the MTF profile. You must know the spot's shape before calculating the MTF. *Figure 6-10* shows the

line spacing obtained with this method for a Gaussian spot shape and an idealized spot shape which plots as a rectangle. In reality, the spot is somewhere between rectangular and Gaussian. Whatever the shape of the current distribution plot, the perceived CRT spot is round.

Design Trade-Offs

You can improve resolution in both refresh-vector and vector-storage displays by trading off brightness for a smaller spot size. Spot size is the limiting factor and lower beam currents produce a smaller spot. However, you can exceed your eye's resolution at normal viewing distances.

Video Amplifiers

These also determine spot size resolution in the horizontal axis. The time between adjacent pixels is:

$$t \text{ (pixel)} = T \text{ (frame)}/N_v \, N_h$$

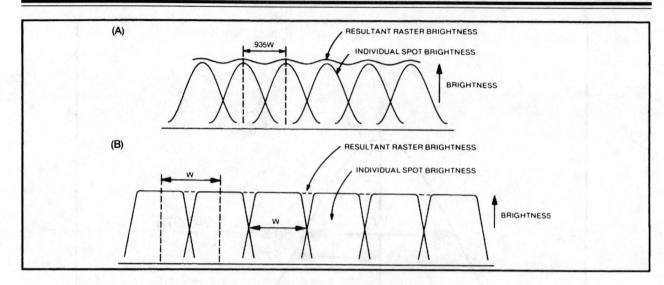

Figure 6-10. *Scan line spacing for a Gaussian spot shape (A) and an idealized (rectangular) spot shape (B).*

T (the frame) is equal to the active frame time. Nv is the number of scan lines, and Nh is the number of pixels per horizontal line. As the number of pixels [Nv x Nh] increases, so does the bandwidth required of the video amplifier. To reduce bandwidth, some raster display systems use a 30 Hz rather than a 60 Hz interlaced raster. This unfortunately increases the tendency to flicker.

Addressability Revisited

As previously stated, addressability is necessary for high resolution and many pixels or vectors, but does not solely guarantee it. Vector displays have inherently high addressabilities, limited by D/A converters. As an example, a vector display capable of resolving 1,000 lines may be able to position vector end points on a 4,000 x 4,000 grid. Generally, to ensure a smooth line, the D/A converters limit addressability to about four times the resolution. In a shadow mask display, the number of resolvable pixels is a function of the screen size. (See *Figure 6-11*)

Raster Display Limitations

Raster displays are more limited due to the fixed pattern of the beam. The number of pixels limits the size of the bit-map's addressability. The num-

ber of scan lines vertically limits addressability and the video resolution horizontally limits it. Increasing the number of pixels in either the vertical or horizontal dimension requires a corresponding increase in the pixel clock rate and a faster bit-map memory system. The number of scan lines and the bandwidth of the video Z-axis amplifier determines useful bit-map size.

Image Artifacts

Vector displays present relatively smooth lines; however, raster displays produce artifacts in addition to the desired image. An artifact is any unwanted image caused by the sampling effects of the raster pattern. Sampling produces high spatial frequencies on the display which do not exist in the intended image. Higher frequency artifacts appear as noise on any edges oriented in any direction other than vertical or horizontal. Aliasing makes a sloping line look like a staircase rather than a smooth line. This staircase is called a "jaggie".

Moire Patterns

Moire is an artifact produced by either the interference between the frequencies of image lines or the interference of the raster lines with the sampling frequency of the shadow mask. When the spacing of raster lines is close to the spacing of the mask, brightness varies periodically across what should be a uniformly colored field unless the spot size exceeds the shadow-mask spacing by a sufficient margin. Generally, you would select a spot size from 1.2 to 1.5 times the shadow-mask pitch, with the raster spacing about equal to the spot width. Resolution depends on spot size, as mentioned previously. Therefore, the shadow-mask pitch is the primary limitation to resolution on a shadow-mask display.

COLOR QUALITY CHARACTERISTICS

The quality of color includes four factors:

1. Brightness.
2. Contrast.
3. Purity.
4. Convergence.

Brightness

The brightness (B) of a CRT is determined by:

$$B = k \, E \, IV/\pi \, A$$

k = the attenuation factor due to the glass faceplate and the shadow mask, E = the phosphor efficiency, I = the time averaging of the beam current, and V = the accelerating voltage in volts. A = the scanned or written-to area.

The Penetron CRT's brightness in a refresh-vector system can be very high. The beam current and the current saturation of the phosphor solely limit it. High beam current also decreases the efficiency, E. Conversely, the DVST with color write through has both limited brightness in the storage and refresh modes. The low-voltage flood electrons exciting the phosphor solely limits the stored-image brightness. The color refresh image brightness is limited, although it is produced by the high-voltage writing beam since the writing beam current must be low to prevent unintentional storage.

The color raster display's brightness, although good for low resolution CRTs, is limited by shadow-mask interception of about 80 percent of the beam current. The presence of three beams partially compensates for this loss.

Contrast

Two mechanisms, intrinsic and extrinsic contrast, limit the CRT display's contrast. Intrinsic contrast is the contrast of the written parts of the image relative to the image's unwritten parts. These are measured in a dark room. The intrinsic contrast, C_i is:

$$C_i = Bw/Bunw$$

Bw and Bunw are the brightnesses of the written-to and unwritten-to areas of the screen respectively. This contrast is quite high for both the refresh-vector and raster color displays. The DVST has low intrinsic contrast because the unwritten to areas of the target receive some excitation from the flood guns. Extrinsic contrast (Cex) much better relates to the "real world" since it accounts for both reflected and scattered light:

$$Cex = (Bw + R)/(Bunw + R)$$

R is reflected and scattered ambient light off the phosphor and screen surface. Other variables are the same as in previous equations. Since all three types of color displays reflect, and in turn scatter approximately the same amount of ambient room light, brightness primarily determines their contrasts. An optical glass filter in front of the screen improves extrinsic contrast. These filters obviously attenuate emitted light while reflected light must make a "double pass" and is therefore attenuated twice.

There are more efficient filters which are more selective, absorbing room light while transmitting the light emitted from the display. The display can have either an "antiglare" coating added to the front or a matte finish to prevent specular reflections.

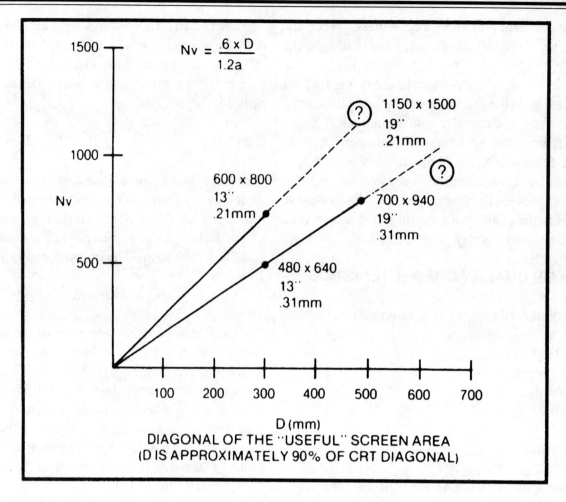

$$Nv = \frac{.6 \times D}{1.2a}$$

1150 x 1500
19"
.21mm

600 x 800
13"
.21mm

700 x 940
19"
.31mm

480 x 640
13"
.31mm

Nv

D (mm)
DIAGONAL OF THE "USEFUL" SCREEN AREA
(D IS APPROXIMATELY 90% OF CRT DIAGONAL)

Figure 6-11. *Screen size versus pixels.*

Color Purity and Convergence

Color purity is the uniformity of color over a large area of the screen. Purity tests measure a color's spectral purity. This is not a problem with the DVST/CWT; however, it can be a severe problem with the shadow-mask color CRT. Their three color beams have to excite the entire phosphor dot — and only that single dot! If the beam spills over into another color beam's place, the color lacks proper clarity. Exact alignment of the electrons passing through the shadow-mask hole would eliminate this concern. However, slight misalignment errors are unavoidable from manufacturing process tolerances. Thermal distortions of heated tubes are also a culprit. Convergence measures if each primary color image is in perfect registration with the other two primary color images. Convergence usually varies as you measure over the CRT.

Chapter 6 Quiz

1. Which of the following is not a color CRT type?
   A. Penetron.
   B. Color write through.
   C. Raster scan.
   D. Shadow mask.
2. Which of the following allows obtaining of different beam Penetron potentials?
   A. Switching target potentials.
   B. Changing CRT colors.
   C. Using electron guns at different potentials.
   D. A and C.
3. The color write through system:
   A. Uses a direct view storage tube.
   B. Has an insulating layer separating the phosphor from a transparent conductor.
   C. Both A and B.
   D. None of the above.
4. Which color tube technology has the most complex CRT?
   A. Shadow mask.
   B. Penetron.
   C. Color write through.
   D. None of the above.
5. Which is not a type of color display system?
   A. The alignment/registration ensuring system.
   B. The vector storage.
   C. The refresh vector.
   D. The raster refresh.
6. A Penetron CRT:
   A. Has red data write to the second field.
   B. Obtains a third color by overlapping but this is difficult.
   C. Write green data in the first field.
   D. Uses non-field-sequential writing field techniques.
7. A vector storage color system:
   A. Is very different than a refresh vector system and its speed is critical with a DVST while in the storage mode.
   B. Solves the flicker imposed limit on the number of vectors.
   C. Both A and B.
   D. None of the above.
8. The refresh vector color system uses:
   A. Two beams deflected together over the phosphor in a predetermined raster pattern.
   B. Three beams deflected together over the phosphor in a predetermined raster pattern.
   C. Has a bit mapped memory which determines the time and magnitude of current of each color gun.
   D. Both B and C.
9. Why is the data in the bit mapped memory read out rapidly?

    A. To avoid misalignment.

    B. To avoid color distortion.

    C. To avoid flicker.

    D. All the above.

10. What affects image quality the most?

    A. The number of scan lines.

    B. Resolution.

    C. Contrast.

    D. All the above.

11. Resolution primarily determines:

    A. The number of scan lines.

    B. Brightness.

    C. An image's sharpness.

    D. Contrast.

12. What primarily determines resolution of a vector type display?

    A. Contrast.

    B. The electron beam spot's size.

    C. The numbers of lines which are equal to the spot's width.

    D. None of the above.

13. Video amplifiers

    A. Help determine spot size in the horizontal axis.

    B. Help determine spot size in the vertical axis.

    C. Help determine spot size in both the vertical and horizontal axes.

    D. None of the above.

14. Artifacts are:

    A. Any unwanted image caused by the sampling effects of the raster pattern in a raster color display system.

    B. Any unwanted image caused by straight flat lines called jaggies.

    C. Caused by the display's sampling induced high spatial frequencies which do not exist in the intended image.

    D. Both A and C.

15. Which of the following is not a factor in determining color quality?

    A. Brightness.

    B. Purity and contrast.

    C. Convergence.

    D. All the above.

Chapter 7
LED Secondary Optics: Accessories, Lenses and Optical Filtering

Chapter 7
LED Secondary Optics: Accessories, Lenses and Optical Filtering

This chapter examines the underlying optics and theory of light filters, followed by some practical filtering techniques and products to enhance light source readability. These filters include special wavelength filters, combination filters, louvered filters, circularly polarized filters, anti-reflection filters, and surface coatings. Collectively, these efforts and products are secondary optics. This also includes backlighting with light pipes, clustered LED arrays, and clustered arrays built into light pipes. This creates new mounting techniques, making front panels more attractive and readable, even in bright ambient light.

LIGHT FILTER THEORY

Contrast and Contrast Ratio

Optical filters have various inherent characteristics which attenuate or diminish the amount of light passing through them. Your immediate reaction may be, why use them if they just cut down on the light output? The answer is contrast enhancement! It maximizes the contrast between the on and off conditions in the display. You achieve this by:

1. Reducing the reflected ambient light on the display.
2. Maximizing the light passing through the filter reaching you.

Contrast and contrast ratio parameters characterize these two stated goals. Contrast varies from 0 to 1 and is a ratio of the luminous sterance to the background luminous sterance. Sterance is the illuminated element's intensity per unit area. Luminous sterance are factors possibly causing loss of detail, rendering subjects indistinguishable from their backgrounds. Contrast ratio is the source luminous sterance divided by the background lumi-

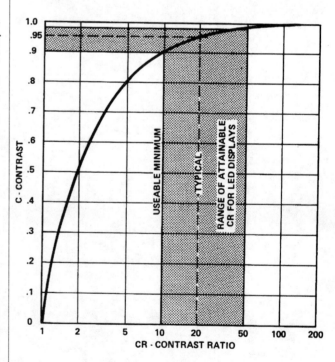

Figure 7-1. *The relationship between attainable C and CR in IED displays. (Courtesy of Hewlett-Packard)*

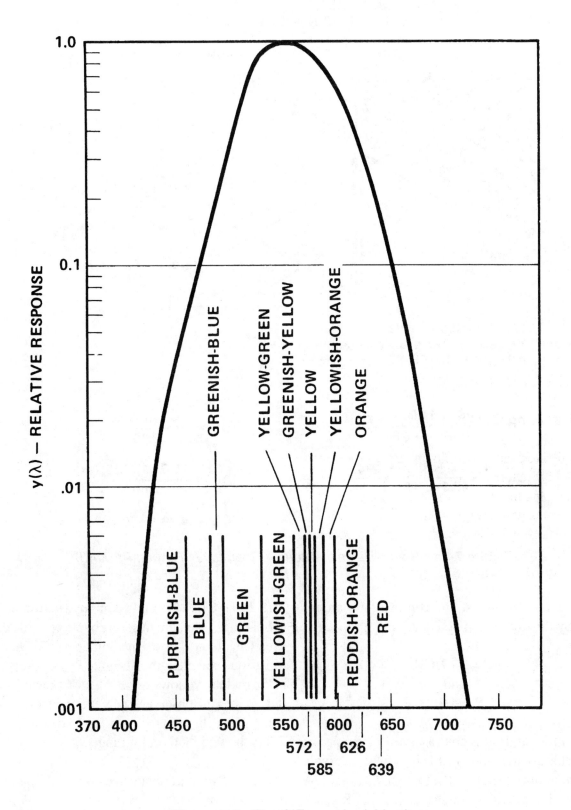

Figure 7-2. *The CIE curve's vivid colors.*

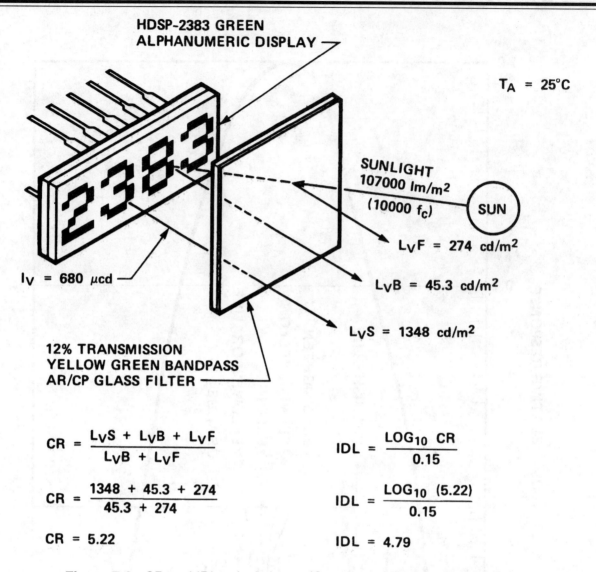

HDSP-2383 GREEN ALPHANUMERIC DISPLAY

$T_A = 25°C$

SUNLIGHT 107000 lm/m² (10000 f_c)

$L_V F = 274$ cd/m²

$L_V B = 45.3$ cd/m²

$L_V S = 1348$ cd/m²

$I_V = 680$ µcd

12% TRANSMISSION YELLOW GREEN BANDPASS AR/CP GLASS FILTER

$$CR = \frac{L_V S + L_V B + L_V F}{L_V B + L_V F}$$

$$CR = \frac{1348 + 45.3 + 274}{45.3 + 274}$$

$$CR = 5.22$$

$$IDL = \frac{LOG_{10} \; CR}{0.15}$$

$$IDL = \frac{LOG_{10} \; (5.22)}{0.15}$$

$$IDL = 4.79$$

***Figure 7-3**. CR and IDL calculations. (Courtesy of Hewlett-Packard)*

nous sterance. Contrast and contrast ratio are related by the following two equations.

$$C = 1 - 1/CR$$
and
$$CR = 1/1 - C$$

C is contrast and CR is contrast ratio. The range of attainable contrast for an LED is between a usable minimum of 0.90 (CR = 10) to 0.98 (CR = 50). A typical filter provides a contrast of 0.95 (CR = 20). Use these equations to confirm this. *Figure 7-1* is the relationship between C and CR attainable for LED displays.

The 1931 CIE (Commission Internationale De L'Eclairage) standard observer or photo-optic chart (See Chapter 1) shows the human eye's response to various wavelengths of light. This curve peaks in the middle (yellowish-green), at 555 nanometers. It describes the observer's perception of an object's color when illuminated by a light source. *Figure 7-2* is the vivid CIE color ranges.

There are two important wavelengths, the *peak* and *dominant* wavelengths. A *peak wavelength* (λ_p) is the wavelength at the peak of the radiated spectrum. Use this to estimate the amount of emitted light which passes through an optical filter. For

EFFECTS OF MAGNIFICATION ON INTENSITY
VIEWING ANGLE, AND APPARENT SIZE

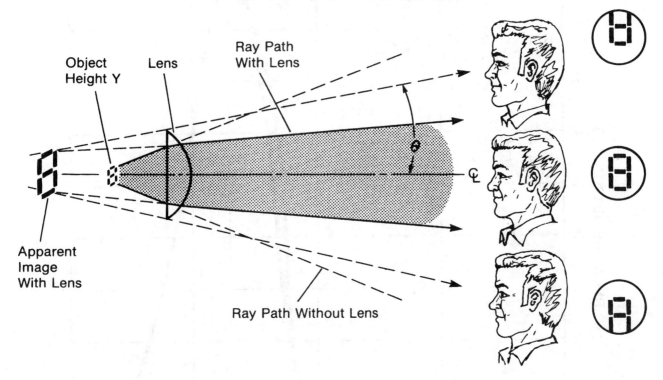

Effects of Magnification: Enlarges Apparent Image Size
Increases Intensity
Limits Viewing Angle

Figure 7-4. *The effects of magnifiers. (Courtesy of Hewlett-Packard)*

example, if a filter has a relative transmission of 40% at a given λ_p, then 40% of the light will pass through and the filter absorbs 60%.

Dominant wavelength, λ_d, for an LED is that portion of the color spectrum which, when mixed with the light from the CIE illuminant C, will be perceived as the same color produced by the radiated spectrum. The CIE illuminant C is a 6,500 K color simulating an overcast day.

The contrast filter must be mechanically stable against temperature expansion and contraction, be chemical resistant, free from visual defects and have a homogeneous index of refraction. That last characteristic defines the amount of incident light the filter/air interface reflects. It is:

$$R = (n_1 - n_2)^2/(n_1 + n_2)^2$$

n_1 = the index of refraction of the filter material and n_2 = the index of refraction of air = 1.

A plastic filter with an index of refraction equal to 1.5 reflects about 4% of the normal incident light at the filter/air interface for a total of 8%. Additional light is absorbed inside the filter by the tinted material. Assuming the optical filter's dye is a constant color, the transmission through this filter material at any given wavelength, T_λ, is a function of the thickness of the material:

$$T_\lambda = e^{-ax}$$

x = the thickness of the filter, e = 2.718, and a = absorption coefficient.

185

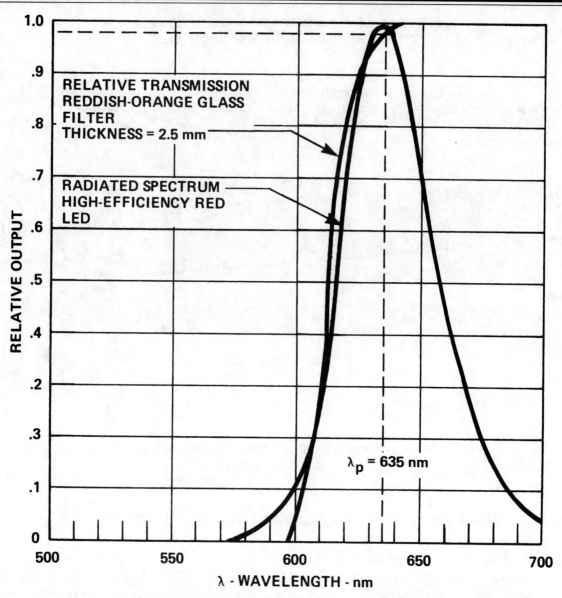

Figure 7-5. *A reddish-orange optical glass filter used with highly efficient LEDs. (Courtesy of Hewlett-Packard)*

Display Illuminance Contrast Ratio

The following equation describes a display without a filter's luminance contrast ratio:

$$CR = L_vS + L_vB + L_vF/L_vB + L_vF$$

L_vS = sterance of illuminated element through the filter, L_vB = sterance of light reflected off the background through the filter, and L_vF = sterance of light reflected off the filter.

Our eyes' response to contrast ratio is the *luminance difference*, EL, described in the next equation:

$$EL = Log\ CR$$

Photography and TV have demonstrated the eye's smallest detectable CR is 1.05. Placing 1.05 CR in the previous equation yields EL = 0.021, which is actually called the threshold luminance difference (ELTH). But for a more comfortable EL, you re-

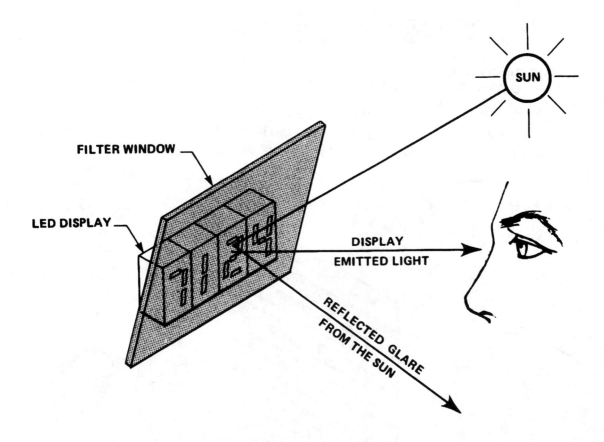

Figure 7-6. *Slanting an optical filter. (Courtesy of Hewlett-Packard)*

quire a CR of 1.4 which gives an EL from the previous equation of 0.15. This EL is the unitary luminance difference, (ELU). The next equation shows that by dividing these quantities you derive the luminance index, IDL:

$$\text{IDL} = \text{EL/ELU} = \text{Log CR/0.15}$$

Let's try a sample calculation. (See *Figure 7-3*) This CR is 5.22 and the IDL is 4.79.

Magnification

Magnification is the ratio of image size to the object's real size. However, despite increasing intensity and apparent size, it limits viewing angle and displaces the image if viewed too far off axis. (See *Figure 7-4*)

ACTUAL LIGHT FILTERS

Plastic Filters

Most manufacturers provide relative transmittance curves for their products. You need to select the total transmittance curve shape and wavelength cutoff points in direct relationship to the LED's radiated spectrum. Three of many manufacturers of plastic filters are: Panelgraphic Corp.'s Chromafilter[tm], SGL's Homalite[tm] and Rohm and Haas Company's Plexiglas[tm]. The LED filters produced by these three companies are usable with all LED display and lamp devices.

Optical Glass Filters

These are designed with constant density so it is their thickness which determines optical density.

**AVAILABLE OPTIONS FOR LOUVERED FILTERS –
ANY COMBINATION IS POSSIBLE**

ASPECT RATIO AND VIEWING ANGLE	LOUVER ANGLE	LOUVER COLOR
2.75: 1 = 60° 2.00: 1 = 90° 3.50: 1 = 48°	0° 18° 30° 45°	OPAQUE BLACK TRANSLUCENT GRAY TRANSPARENT BLACK

EXAMPLE: 2.75: 1 – 18° – TRANSPARENT BLACK

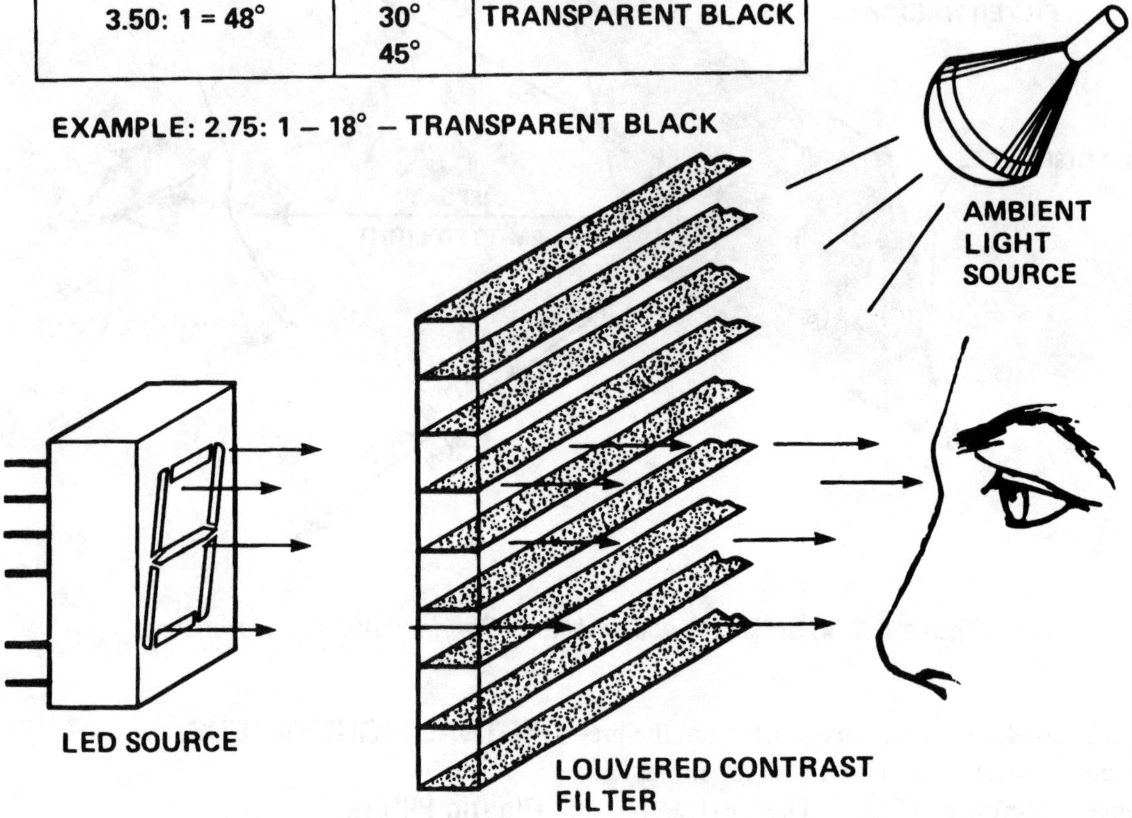

Figure 7-7. A louvered filter's aspect ratio. (Courtesy of Hewlett-Packard)

This is just the opposite of plastic filters which are usually designed so all material thicknesses have the same optical density. Glass filters have superior performance (especially with red LEDs) over plastic filters but cost far more.

The all important relative transmittance is higher with glass and the curve is far steeper and more closely resembles that of the red LED itself. This filter also provides excellent contrast enhancement in harsh ambient light. *Figure 7-5* is a reddish-orange optical glass filter suitable for a high-efficiency red display in moderate ambient light.

Combination Filters

You can use a neutral gray filter with other filters to provide a gray filter window and increase contrast in bright ambient conditions. *Figure 7-6* shows slanting a filter away from the display's flush surface, to direct reflections away from your eyes. The relative transmittance is equal to the product of each filter. The advantage is a dark gray front panel window with very low luminous sterance (zero transmission below 525 nm) which retains its appearance in bright ambient conditions. The tradeoff is a reduced contrast ratio. This is offset though by

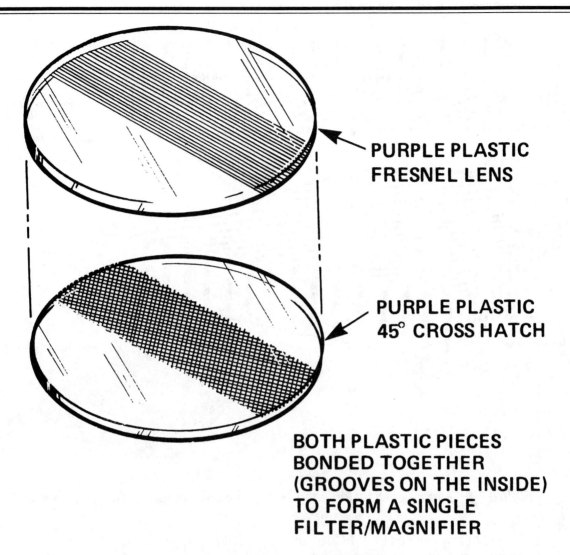

PURPLE PLASTIC FRESNEL LENS

PURPLE PLASTIC 45° CROSS HATCH

BOTH PLASTIC PIECES BONDED TOGETHER (GROOVES ON THE INSIDE) TO FORM A SINGLE FILTER/MAGNIFIER

Figure 7-8. A Fresnel lens magnifier cross-hatch purple filter for red LEDs. (Courtesy of Hewlett-Packard)

the distinct color difference between the illuminated yellow segments of the display and the dark gray background.

Louvered Filters

Louvered filters reduce the display's bright reflected light. They do this without substantially reducing the display's emitted light. *Figure 7-7* shows the louver aspect ratio (the louver depth/distance between adjacent louvers) versus the allowable viewing angle. You have two options with these devices. First, there is a 45° neutral density filter producing a 60° viewing angle with a louver aspect ratio of 2.75. Secondly, a crosshatch filter-

ing scheme (See *Figure 7-8*) reduces the viewing angle to just 40°, but increases contrast. The purple crosshatch filter is a mixture of blue and red. Blue is a psychologically pleasing clearly distinct background color for red.

Figure 7-9 is a more effective optical filter, providing contrast in bright sunlight by using a 45° louvered filter. The difference in this and an ordinary crosshatch louvered filter is the louvers are transparent black with a cross light transmission of 12% to 15%. This small amount of cross transmission helps eliminate ghosting problems of earlier designs.

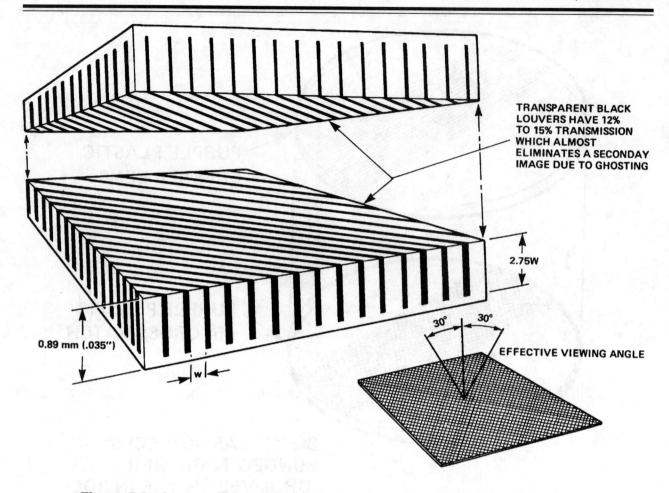

TRANSPARENT BLACK LOUVERS HAVE 12% TO 15% TRANSMISSION WHICH ALMOST ELIMINATES A SECONDAY IMAGE DUE TO GHOSTING

2.75W

0.89 mm (.035")

w

30° 30°

EFFECTIVE VIEWING ANGLE

Figure 7-9. A more efficient louvered optical filter. (Courtesy of Hewlett-Packard)

Louvered filters are available from 3M Corp. under the trade name "Light Control Film™." Louvered filters come with either a "light matte" or "very light matte" anti-reflection surface which does not produce a fuzziness in the display's appearance. In contrast, an ordinary textured anti-reflective filter may reduce readability in direct sunlight. (See *Figure 7-10*) Another option is a hard scratch resistant surface.

Circular Polarizing Filters

These type filters are effective with specular reflecting surfaces. Solid reflecting surfaces with incident light arriving at angles less than 90° reflect light equally in all directions, without scattering. Displays with polished glass or plastic facial surfaces fit this category, like *Figure 7-11*, and are made of a laminate (layer) of linear polarizing material, a quarter wave plate and a circular polarizer. Its optical axis is parallel to the polarizer's flat surface and 45° to the linear polarization axis.

The advantage of a circular polarizer is it reduces ambient light by more than 95%. However, the light passing through this filter is unfortunately attenuated by 65%. Circular polarizers are normally colored to obtain additional wavelength filtering. If you use one of these filters outside though you'll need an ultraviolet filter in front of the polarizer.

SPECIAL LIGHT FILTER END PRODUCTS

Night Vision Goggles

One practical example of effective optical filtering is the military's night vision goggles. The avail-

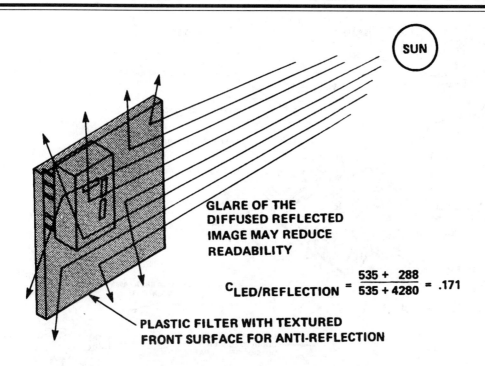

GLARE OF THE
DIFFUSED REFLECTED
IMAGE MAY REDUCE
READABILITY

$$C_{LED/REFLECTION} = \frac{535 + 288}{535 + 4280} = .171$$

PLASTIC FILTER WITH TEXTURED
FRONT SURFACE FOR ANTI-REFLECTION

Figure 7-10. An anti-reflective filter. (Courtesy of Hewlett-Packard)

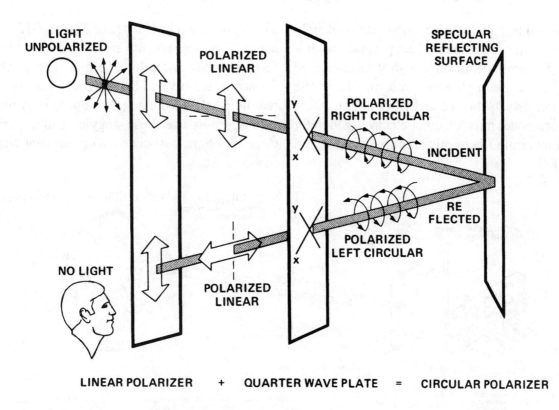

LINEAR POLARIZER   +   QUARTER WAVE PLATE   =   CIRCULAR POLARIZER

Figure 7-11. Antireflective circular polarization. (Courtesy of Hewlett-Packard)

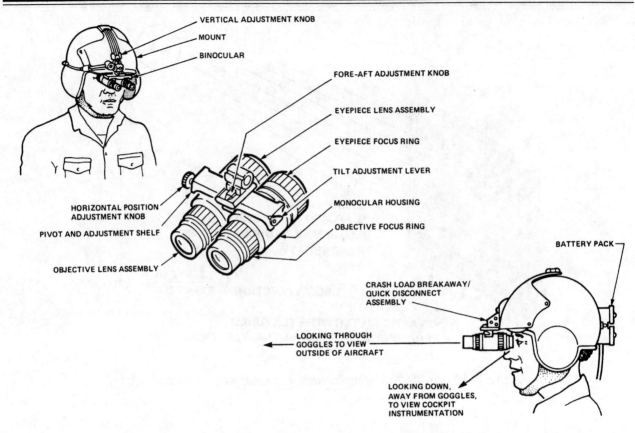

Figure 7-12. *Night vision goggles. (Courtesy of Hewlett-Packard)*

able red and IR light from the moon, stars and night sky, once sufficiently intensified, produces an image in moderate detail on a highly miniaturized monochrome green phosphor screen. The image intensifier detects this red and IR radiation, using a photocathode, microchannel electron generator and an inverting fiber optic amplifier with a gain of 10,000.

The night vision goggles (See *Figure 7-12*) weigh about 20 ounces and are small binocular type devices attached to a pilot's helmet. The quick disconnect assembly and battery pack at the helmet's rear help balance the goggles. Focus is by manual means. Pilots use these goggles and experience three problems associated with ambient lighting in a cockpit:

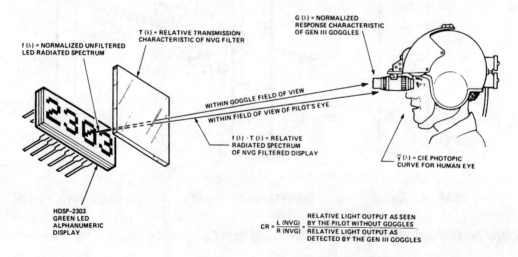

Figure 7-13. *A night vision goggle's figure of merit. (Courtesy of Hewlett-Packard)*

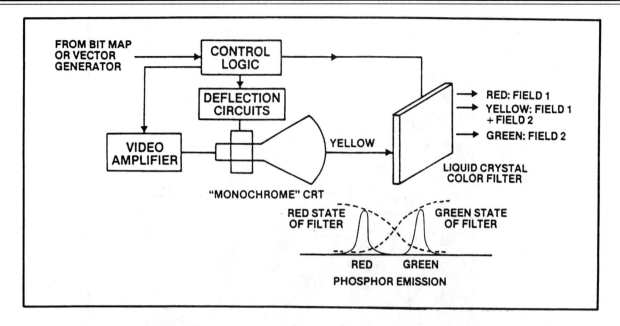

Figure 7-14. *The liquid crystal shutter display. (Courtesy of Hewlett-Packard)*

1. Direct light causes blooming or goggle shut down.
2. A source of light reflected off the windshield causes contrast loss or hinders definitive object recognition.
3. A light source not in the field of view, nor reflecting, produces sufficient light to flare and scatter light.

Figure of Merit

This is the compatibility ratio for night vision goggles. It is the ratio of the relative light output as seen without goggles to the relative light as detected with the night vision goggles. This is again the CR ratio. (See *Figure 7-13*)

Helping Factors

Dimming green LED displays enhances night vision compatibility. You do this by either PWM (pulse width modulation) or decreasing the display's current with a series resistor. Green is in the middle of the goggle's perceived spectrum. For a strobed LED display, such as the HDSP-2303 from Hewlett-Packard, PWM is the better technique. Using PWM dimming reduces LED illumi-

nation ON-time (without affecting the strobe rate). You can achieve three orders of magnitude of dimming. If you prefer to dim the display by reducing DC forward current, the current ratio is a 10:1 maximum.

Night Vision Filters

WAMCO (See the list of suppliers - Appendix B) specially designs glass optical filters for night vision goggles. These are for green LED displays and severely attenuate wavelengths of light which reside in the yellow, red and IR regions. There are now commercially available night vision monoculars for slightly over $200.

THE LIQUID CRYSTAL SHUTTER DISPLAY

Changing Monochrome CRTs to Color

Liquid crystal shutter displays yield full field sequential color. This liquid crystal technology offers a number of attractive features in comparison to more conventional shadow-mask color CRTs, for example. Liquid crystal shutters lack screen patterning but yield a better defined smooth continuous non-jagged image. The screen's geometry

193

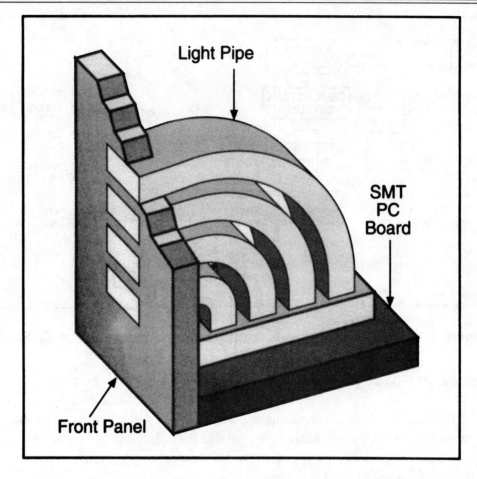

Light Pipe

SMT PC Board

Front Panel

Figure 7-15. *An illustration of light pipes. (Courtesy of Dialight)*

allows a wide range of sizes without the constraints of a certain width-to-height aspect ratio, which is common to all other color displays. This makes it ideal for desktop computers or instrumentation equipment panels. (See *Figure 7-14*)

A History of Frame Sequential Color Displays

Historically, frame sequential color was first used by Baird in England in 1928. This rotating filter wheel simultaneously used a TV camera and monitor. Size constraints made this technique impractical. No progress occurred in viewability constraints for frame sequential color until this technology was revisited with more capability. But problems persisted along the way.

These liquid crystal approaches included twisted nematics (TN), tunable birefringence, and dual fre-

quency addressed nematic liquid crystals. The TN technology lacked sufficient speed, the tunable birefringence method had too restrictive of a viewing angle, and the dual frequency addressed nematic liquid crystals suffered from excessive resistive losses in their transparent conductive coatings, preventing a diagonal larger than 7 inches. Tektronix finally succeeded in developing a LCS display which fit over a monochrome oscilloscope CRT, transforming it into a multi-color CRT.

SECONDARY OPTICS

Light Pipes

LEDs often can't carry light to a desired location. But light pipes (See *Figure 7-15*) transfer generated light from one place to another. Traditional light pipes are attached to the front panel or to a circuit board because they can't survive the heat

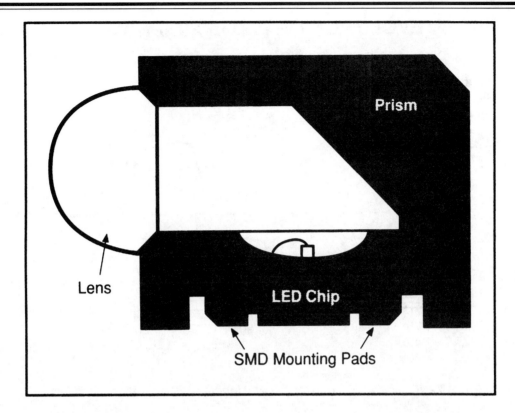

Figure 7-16. *The Dialight Prism^tm light pipe. (Courtesy of Dialight)*

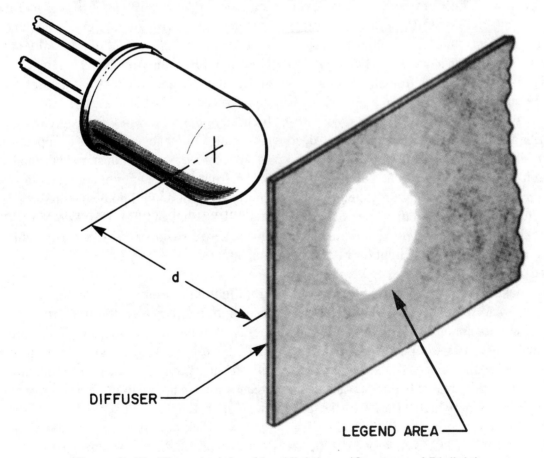

Figure 7-17. *The principle of backlighting. (Courtesy of Dialight)*

EVOLUTION OF LED EFFICIENCY

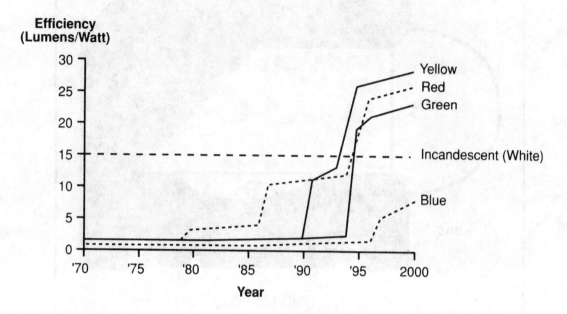

Figure 7-18. *The evolution of LED efficiencies. (Courtesy of Dialight)*

of surface mounting components. However, Dialight's new PRISM CBI™ series light pipes withstand this temperature. The Dialight Prism uses a prismatic light pipe in a housing. (See *Figure 7-16*) This refracts the light 90° from the vertical axis to the viewing area.

Backlighting Ideally, a front panel has a "dead front" appearance for all symbols, except those which aren't supposed to be seen. You do this by making them visible with light projected through from a light source in back of the panel, hence the term "backlighting." (See *Figure 7-17*) Early red LEDs did not produce enough light for backlighting. (See *Figure 7-18*) Recent LED technology advances make this not only possible with red LEDs, but now accommodate inherently less efficient yellow and green LEDs. This technique makes panels appear smooth and finished. Special purpose LEDs for this task (See *Figures 7-19a* to *7-19d*) usually reside behind a translucent plastic (mylar) window which is flush with the faceplate. Alternately, LEDs directly illuminate the legend. (See *Figure 7-20*) Crosstalk, or light bleed-

ing from adjacent diffused LEDs sometimes occurs and spreads (diffuses) light. Note the ON and OFF LEDs in *Figure 7-21* don't totally appear as ON and OFF on the front panel legend of the male and female symbols. Crosstalk also occurs with non-diffused LEDs, but to a lesser extent since their focused light on the window has greater intensity. Crosstalk can make an unlit LED appear on from an adjacent LED's light transmitted through its lens to the front panel. Dialight addresses this with a family of right-angled housings or shrouds completely covering an LED. (See *Figure 7-22*) These ensure zero crosstalk and extremely bright on-axis indicators.

Frontlighting substitutes for or supplements ambient lighting; therefore, it differs from backlighting. Backlit small symbols, such as a rectangular LED, have built-in diffusants for uniformity requirements. Using a cluster to illuminate a large area is more an application for a contrast enhancement filter. If you want separate illumination areas, use black tape separators to avoid crosstalk. You can backlight with undiffused encapsulated

(a) Square LED backlighting displays.

(b) Horizontally stackable light module backlighting displays.

(c) Various sizes and shapes of backlighting bargraph stackable modules.

(d) Examples of backlit front panels.

Figure 7-19. *Special purpose LEDs. (Courtesy of Hewlett-Packard)*

LEDs. (See *Figure 7-17* again) They yield fairly sharp edges when the distance "d" is small. *Figure 7-24* shows the shortest distance possible behind the panel for backlighting use of an "egg-crate" reflector. This shows how to place the diffuser, legend plate, and (if used) contrast enhancement filter. Light pipes also provide indirect illumination. This plastic is usually a solid piece with a hole to receive the LED. It relies on greater than critical angle reflections for its efficiency. Scratches and

contact with materials other than air cause light losses at the surface; therefore, the surface must be protected.

If you use this technique over a length of four diameters, multiple reflections may cause enough diffusion to allow you to eliminate the separate diffuser. (See *Figure 7-23*) Making "d" smaller results in a sharper front panel light pattern. You can also roughen the front end of this type of plastic

Figure 7-20. *An LED directly contacting and illuminating a legend. (Courtesy of Hewlett-Packard)*

light pipe if the length is marginal. When back-lighting larger areas, you can obtain maximum efficiency by seating each LED in an "egg-crate" type reflector in a honeycomb or square configuration. (See *Figure 7-24*) In this illustration there are an equal number of LEDs but a different aspect (height-to-length) ratio. The honeycomb arrangement has a small advantage of 2 divided by the square root of 3 (1.155) over the square configuration alone in light output efficiency. You could manually configure these LED clusters; however, LEDs now are already formed into PC board mountable clusters.

Figure 7-21. *The light bleeding principle. (Courtesy of Hewlett-Packard)*

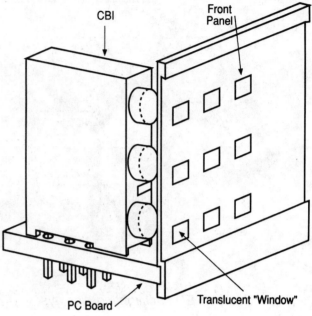

Figure 7-22. *Right angle light pipes. (Courtesy of Hewlett-Packard)*

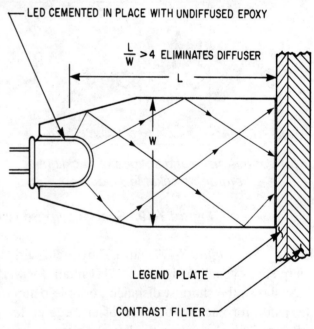

Figure 7-23. *Eliminating backlighting diffusers. (Courtesy of Hewlett-Packard)*

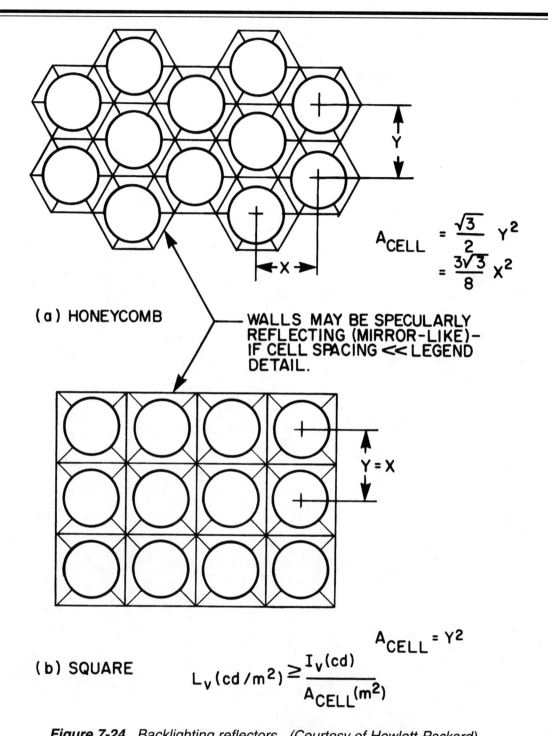

$$A_{CELL} = \frac{\sqrt{3}}{2} Y^2$$
$$= \frac{3\sqrt{3}}{8} X^2$$

(a) HONEYCOMB

WALLS MAY BE SPECULARLY REFLECTING (MIRROR-LIKE)— IF CELL SPACING ≪ LEGEND DETAIL.

$$Y = X$$

$$A_{CELL} = Y^2$$

(b) SQUARE

$$L_v \,(cd/m^2) \geq \frac{I_v \,(cd)}{A_{CELL} \,(m^2)}$$

Figure 7-24. Backlighting reflectors. *(Courtesy of Hewlett-Packard)*

Quiz Chapter 7

1. Techniques and products designed to improve the readability of optoelectronic devices are collectively called:
   A. Filters.
   B. Secondary optics.
   C. Polarizers.
   D. Light pipes.

2. When made, light pipes formerly had to be attached to front panels or PC boards because:
   A. Their fibers were too short.
   B. They became clouded.
   C. They couldn't withstand the heat associated with the surface mounting process.
   D. They were too expensive.

3. The backlighting technique suffers from crosstalk or:
   A. Distortion of colors.
   B. A co-mingling of the light of adjacent diffused LEDs.
   C. Making LEDs appear ON when they are really OFF.
   D. B and C.

4. Light pipes are a solid piece of plastic with a hole in them which receive the LED and rely upon:
   A. Fast switching.
   B. Greater than critical angle reflections for their efficiency.
   C. Uninterrupted power.
   D. None of the above.

5. You may be able to _____ if you use the light piping technique at distances greater than a length of four diameters.
   A. Eliminate the separate diffuser.
   B. Condense light.
   C. Conserve light.
   D. B and C.

6. Some filters attenuate light output, but their redeeming quality is contrast enhancement, T or F?

7. The two wavelengths most crucial to you, the observer, are the peak and dominant wavelengths, T or F?

8. Assuming an optical filter's dye is a constant color, the transmission through this filter material is a function of:
   A. Its thickness.
   B. Its diameter.
   C. Its weight.
   D. Its ambient lighting conditions.

9. Louvered filters are most effective at:
   A. Color enhancement.
   B. Keeping the surface clean.
   C. Reducing the amount of light reflected from the display.
   D. Color fidelity.

10. Night vision goggles work best without the yellow, red and IR light associated with _____ LEDs.
    A. Red.
    B. Yellow.
    C. Green.
    D. None of the above.

Chapter 8

Basic Optical Math and Concepts of Optics

Chapter 8
Basic Optical Math
and Concepts of Optics

This brief, profusely illustrated chapter compactly condenses optoelectronics' optical aspects with primary emphasis on lenses, their properties and physical arrangements. The chapter also covers optical glass coatings, beam splitters such as prisms and diffraction gratings and reflectors, mainly mirrors. Mass produced commercially available optoelectronic products have to be inexpensive; therefore, they simply can't use optical quality glass lenses. However, in an optoelectronics system or application, you may well have a stringent requirement which you can only meet with an optical quality lens or lens pair. Therefore, we'll first review some basic optical relationships, definitions, equations and concepts which govern lenses and their uses.

The Basics of a Lens

It is so fundamentally simple and obvious that it almost defeats the description, but a lens, like a window, is a transparent medium through which light passes virtually unattenuated. However, it differs from glass because it focuses (converges) or diffuses and spreads (diverges) light. The most admirable quality of a lens is its high transmittance or ability to reflect only a minute portion of the light back to the source. It admirably passes the vast majority of its light. Lenses pass light so well they have a transmittance approaching 1 and a reflectance approaching 0; therefore, using 1 as the transmittance in equations greatly simplifies calculations. However, typical transmittance ranges for lenses are really 0.95 to 0.99.

Optical Formulae

Figure 8-1 is a basic biconvex lens which means it is thicker in the middle than at its sides. You use lenses like this in magnifiers and some light condensing or concentrating applications. This lens has an *O* to the far left and an *I* to the far right, in *Fig-*

ure 8-1. These represent the *Object* and the *Image*. *Table 8-1* defines all the other variables, except *M*, which is magnification. But let's first define focal length so you can derive the remaining formulae from it. The focal length is the distance from the center of the lens to the point on the image's side where the light converges into a single point. If this sounds abstract, the most practical example demonstrating this is a magnifying glass gathering the sun's rays and concentrating them on a single point which heats to an extent it can produce a fire. The distance from the center of the lens (not its surface) to this smallest possible (and hottest) converging point is the focal length (F) of a lens.

The EFL or effective focal length of a lens assumes that if the light source is far enough away the lens intercepts a number of essentially parallel rays from this source. It then refracts and focuses them to a single point. (See *Figure 8-2*) Note this EFL distance is from the second lens' refracting surface, not its center. The EFL and focal length are very nearly equal in most applications. Stated in another way, it is the distance from the second principal

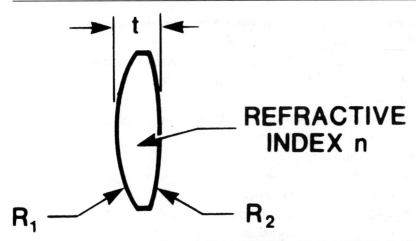

REFRACTIVE INDEX n

R_2 IS NEGATIVE FOR THE CURVATURE SHOWN

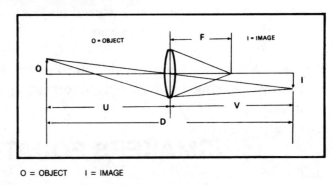

O = OBJECT   I = IMAGE

Figure 8-1. A simple single lens with its basis for lens formulae.

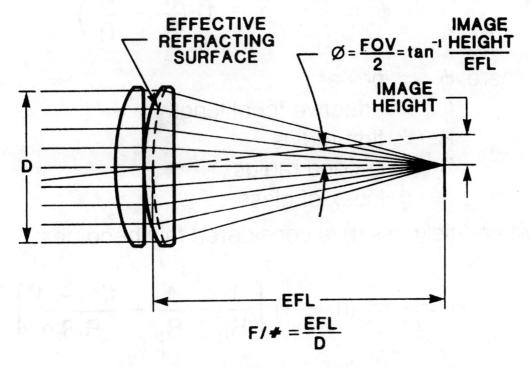

EFFECTIVE REFRACTING SURFACE

$$\emptyset = \frac{FOV}{2} = \tan^{-1}\frac{IMAGE\ HEIGHT}{EFL}$$

IMAGE HEIGHT

$$F/\# = \frac{EFL}{D}$$

Figure 8-2. The EFL (Effective Focal Length) of a lens.

point and the second focal point with parallel incident light. The relative aperture number of a lens is the ratio of its EFL to the diameter of the clear aperture of the lens. This parameter defines the amount of illumination the lens can pass; therefore, this is a critical parameter in photography. Another critical lens parameter is its diopter. This is the reciprocal of the lens' EFL in meters and defines a lens' power. Your eye glasses use this parameter. The Lensmaker's equation uses this concept with f representing power. (See *Figure 8-3*)

Lens Types

There are three lens types which depend on lens groupings. (See *Figure 8-4*) These are:

1. A simple lens which are a single optical element which refracts incident light to form an image.

2. A compound lens which is a group of two or more simple lenses.
3. A complex lens which is composed of a multiple group of lens elements.

There are several common simple lens types in use today which include:

1. Concave.
2. Plano-convex.
3. Biconvex.
4. Cylindrical.
5. Fresnel lenses.

A concave lens is thinner in the middle than at its sides and diffuses rather than concentrates light. It is just the opposite of a convex lens in these qualities. It is somewhat ineffective unless it is part of a lens pair or configuration. A plano-convex lens is seen in *Figure 8-6*. *Plano* means flat and this lens

LENSMAKERS EQUATION

$rA = \dfrac{EFL}{\phi_{CLEAR}}$

$D = \dfrac{1}{EFL}$

$$\phi = \frac{1}{f} = (n - 1)\left(\frac{1}{R_1} - \frac{1}{R_2}\right)$$

where: ϕ = power
f = effective focal length
R_1 = first radius
R_2 = second radius
n = index of glass

when thickness (t) is considered this becomes:

$$\phi = \frac{1}{f} = (n - 1)\left[\frac{1}{R_1} - \frac{1}{R_2} + \frac{t(n - 1)}{R_1 R_2 n}\right]$$

Figure 8-3. Lens makers' equation.

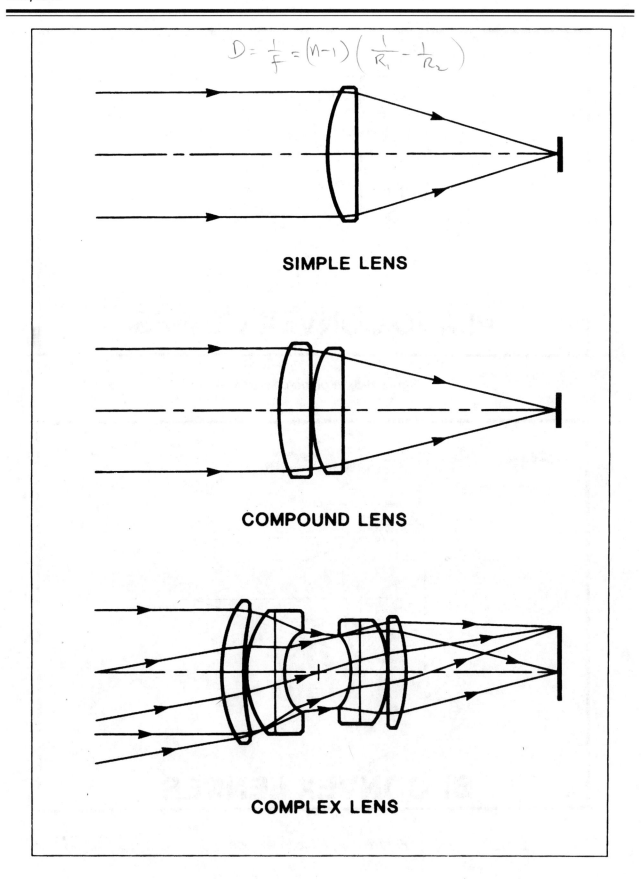

$$D = \frac{1}{F} = (N-1)\left(\frac{1}{R_1} - \frac{1}{R_2}\right)$$

SIMPLE LENS

COMPOUND LENS

COMPLEX LENS

Figure 8-4. *Lens types by grouping(s).*

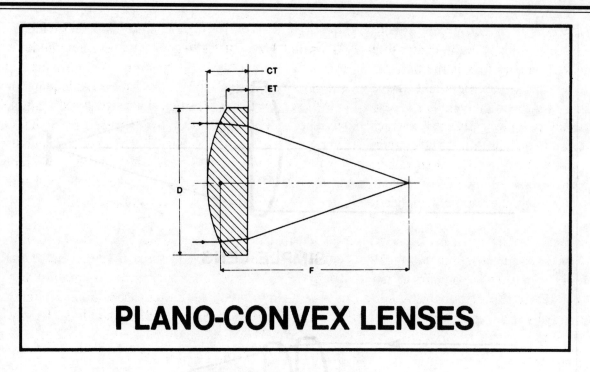

PLANO-CONVEX LENSES

Figure 8-5. *A plano-convex lens.*

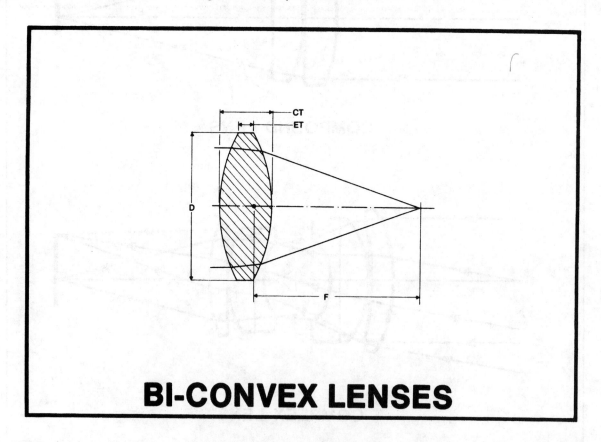

BI-CONVEX LENSES

Figure 8-6. *A biconvex lens.*

has a positive or convergent effect with one flat side and is thicker in the center than at its sides. This condensing lens is the basis of inexpensive microscopes, magnifiers and eyepieces jewelers might use. The biconvex lens (See *Figure 8-6*) is typically symmetrically convex which means it has equal radii on both sides. The cylindrical lens (See *Figure 8-7*) has a focal line, rather than a focal length. This property makes it ideal as an image slit or in other narrow field applications.

A very common former use for a cylindrical lens was on the sliding glass magnifying cursors of slide rules. These were the precursors (so to speak) of today's scientific calculator. A slide rule had a horizontally sliding strip that was secured by being wedged in a tongue-in-grove fashion, between the upper and lower stationary scales. This cylindrical lens cursor was only an inch or less wide, and slid horizontally to magnify everything under its fine center line. This cursor had no more than a tenth-of-an-inch focal line. This was adequate, though, for that application since it required absolutely no *depth-of-field*; it only focused on a flat surface directly beneath it. This fixed or constant distance from the cursor's cylindrical lens' center never varied — the distance could not vary since a cylindrical lens has no focal length!

The Fresnel lens (See *Figure 8-8*) is a more specialized simple lens, like the cylindrical lens. It has a series of concentric grooves on a flat thin piece of plastic. This makes a large diameter, lightweight inexpensive lens with a short focal length. Each groove acts as a minute refracting facet or surface capable of bending light. This yields greater trans-

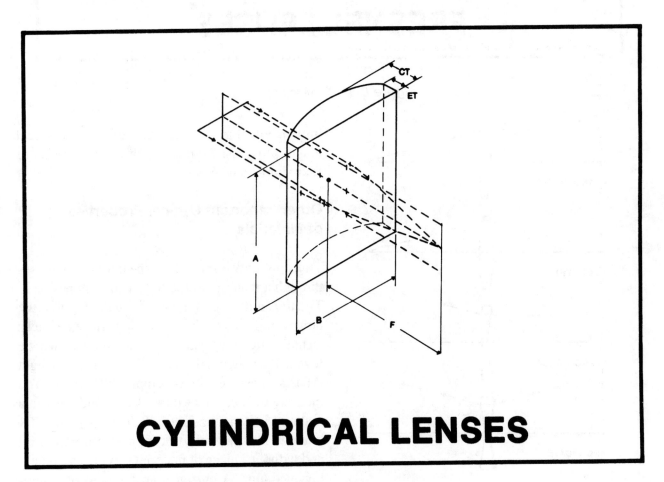

CYLINDRICAL LENSES

Figure 8-7. A cylindrical lens.

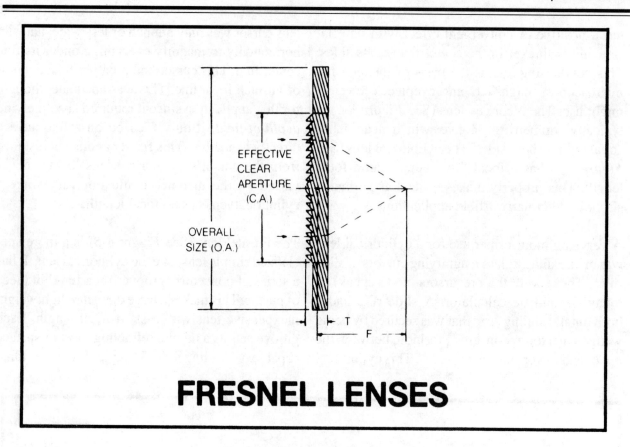

EFFECTIVE
CLEAR
APERTURE
(C.A.)

OVERALL
SIZE (O.A.)

T

F

FRESNEL LENSES

Figure 8-8. *A Fresnel lens.*

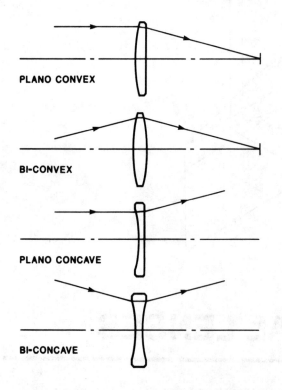

PLANO CONVEX

BI-CONVEX

PLANO CONCAVE

BI-CONCAVE

Figure 8-9. *The effects of four simple single lenses.*

mittance than an ordinary lens of the same focal length provides. *Figure 8-9* shows the effects of four simple lens types.

Other Important Optical Properties of Materials

Transmissivity is the ratio of the light which passes through a material to the light which shines on it. The transmittance of a lens is usually 0.9 to 0.95; however, if the lens has multiple surfaces or lacks an anti-reflective coating, its transmittance will be lower. If a material passes 40% of its incident light it has a transmissivity of simply 40% or 0.4. You measure the detector's current with and without a material is its path. *Figure 8-10* shows the CTR (*current transfer ratio*) of a phototransistor with and without a piece of material serving as a blocking medium. A current transfer ratio of 1 indicates a totally clear material and a ratio of 0 indicates the material blocks all light and is opaque.

SOURCE   DETECTOR

Reference Level is Established with Only Air as Medium

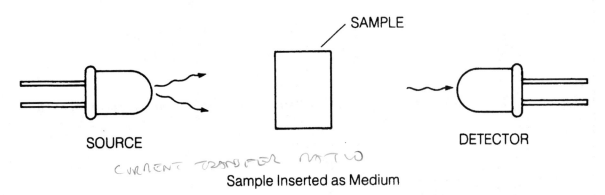

SAMPLE

SOURCE   DETECTOR

CURRENT TRANSFER RATIO

Sample Inserted as Medium

Figure 8-10. *How transmissivity affects the CTR of an opto-emitter and optodetector pair.*

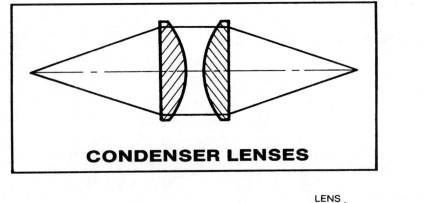

CONDENSER LENSES

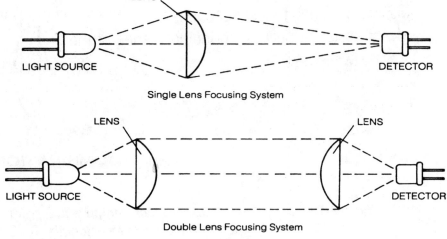

LENS

LIGHT SOURCE   DETECTOR

Single Lens Focusing System

LENS   LENS

LIGHT SOURCE   DETECTOR

Double Lens Focusing System

Figure 8-11. *An optical condensing effect.*

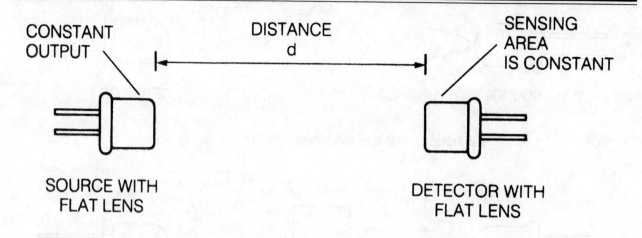

Figure 8-12. *The distance relationship between an opto-emitter and optodetector pair.*

$$\frac{1}{FC} = \frac{1}{F1} + \frac{1}{F2} - \frac{d}{F1 \times F2}$$

or:

$$FC = \frac{F1 \times F2}{F1 + F2 - d}$$

and:

$$d = (F_1 + F_2) - \frac{F_1 \times F_2}{FC}$$

where: FC = focal length of combination
F1 = focal length of first lens
F2 = focal length of second lens
d = distance between **principal** planes of the
two lenses.

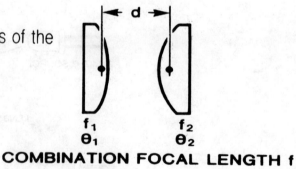

COMBINATION FOCAL LENGTH f

Figure 8-13. *Obtaining the desired focal length with a pair of convex lenses, each with their own focal lengths.*

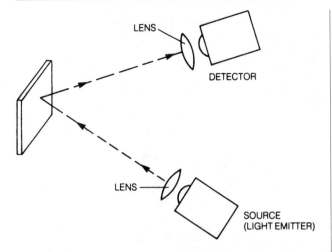

Figure 8-14. *Using mirrors and lenses together.*

This preceding illustration does not take cognizance of the possible condensing effect a material formed into a lens may have which is its primary function in optoelectronics. *Figure 8-11* shows this condensing effect and the shape and type lenses which produce it.

The Relationship Between the Source and Emitter's Distance

As our formulae show, another important optical aspect is the distance between a photodetector and emitter. In the most typical application, you place a flat lens with a constant intensity, wide radiation pattern light source directly in front of an optoelectronics detector. The following equation approximates this distance versus intensity relationship with the light source uniformly emitting light in all directions:

$$K = E_e \cdot d^2$$

FOCAL LENGTH OF SPHERICAL MIRROR

$$f = \frac{r}{2}$$

Where:
   r = Radius of curvature of mirror

E_e is the intensity in watts/meter², d is in inches. The K factor is related to E_e and l by the following equation, if l is one inch (See *Figure 8-12*):

$$E_e \cdot d^2 = 1 \text{ or } d = (1/E_e)^{1/2}$$

Lens Pairs

Since we predominately condense and focus light, *Figure 8-13* describes how you can obtain your desired focal length with a pair of convex lenses, each with their own focal lengths.

Mirrors

A mirror is sometimes less expensive than a single lens and does not pass but rather reflects light back to a detector, for example. The curvature of the mirror is naturally crucial and the smoother, more polished the surface, the more light reflects back to the detector. You may use mirrors in conjunction with lenses. (See *Figure 8-14*) Alignment and focusing then becomes more critical as you increase the sheer number of optical elements. A spherical mirror (See *Figure 8-15*) has a focal length which is a function of its radius of :.

Mirror Precautions

Mirrors are inherently efficient unless there is a dull film on their surface. Mirrors in satellite optical applications therefore have to be wary that no plastics or other materials which "outgas" come within their vicinity. The outgassing phenomenon is when a newer piece of plastic gives off gases which deposit thin films on nearby surfaces. The

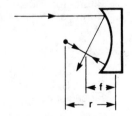

Figure 8-15. *A spherical mirror's focal length.*

Type		How it operates
Cube Polarizers	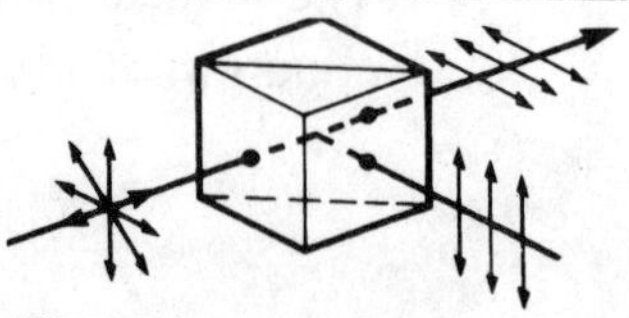	Polarization dependence of thin film reflectance
Uncoated Windows or Wedges	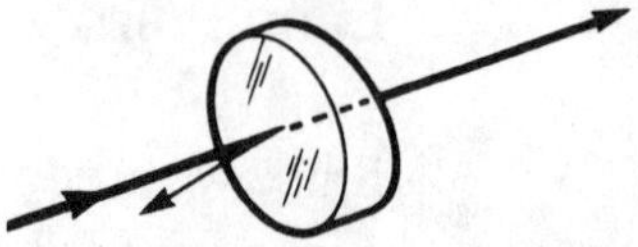	Fresnel reflection Typically 10/90 R/T
Inconel Coated Windows	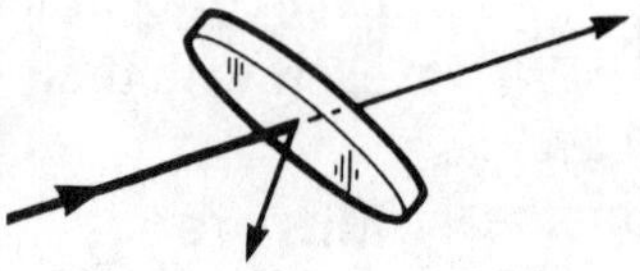	Metallic Reflection 32/32 R/T
Dielectric Coated Window or Wedges	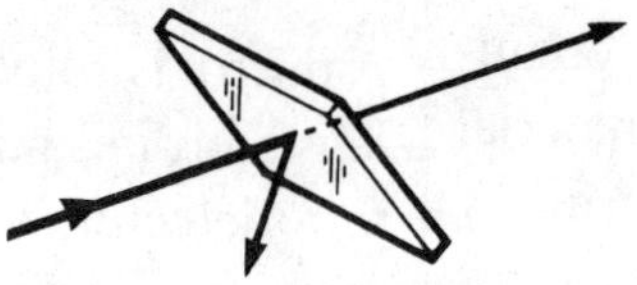	Thin film reflection Various R/T ratios possible
Polka Dot Windows	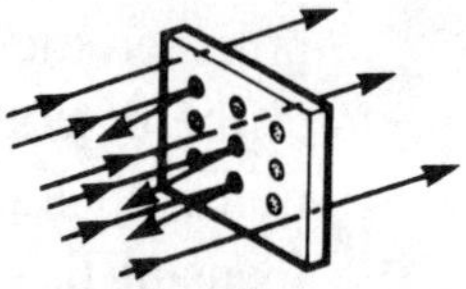	Patterned coating on substrate Spatial: reflecting dots
Cube Beam Splitters	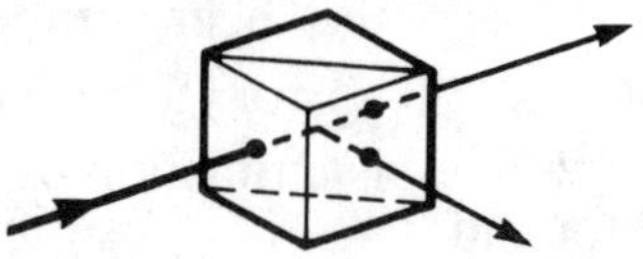	Thin film reflection 50/50 R/T
Pellicles Uncoated or coated	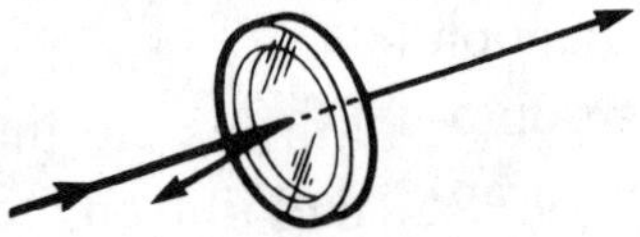	Fresnel reflection Thin film reflection
Grating Beam Splitters	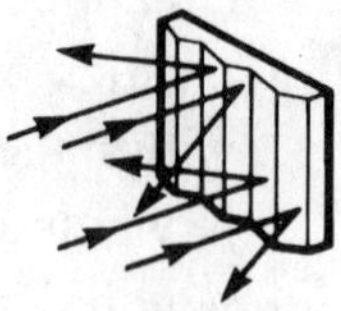	Spatial: inclination of reflecting facets
Bifurcated Fiber Optic Bundles	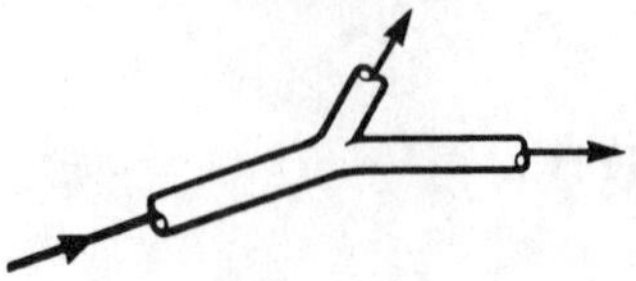	Fiber bundles split into two branches
Beam Splitter Module for Fiber Bundles	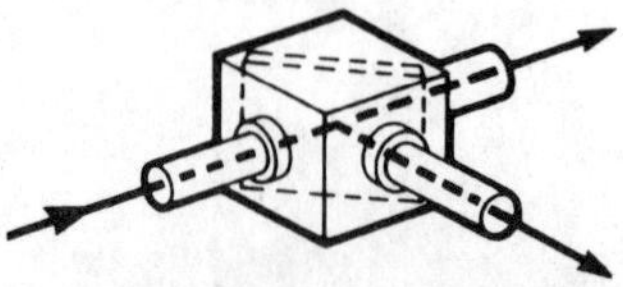	Inconel coated beam splitter (32/32 R/T)

Figure 8-16. Numerous beam splitter types. (Courtesy of Oriel Corporation)

214

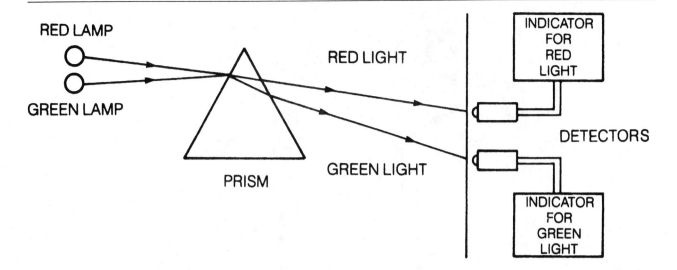

Figure 8-17. *A prism's ability to separate colors.*

best practical example is the "new car's" smell, especially a new car with plastic covers still on the seats. After several days you can notice a film build up on the interior windshield. This is from the plastic covers outgassing.

Prisms and Diffraction Gratings

Together, these constitute a group of optics called beam turners or splitters. A beam splitter splits a beam into two or more separate rays. Depending on the type splitter you use, you may suffer some light radiation absorption. *Figure 8-16* shows numerous beam splitters, some with their R/T or reflectance transmission ratios, as well as their principles of operation. Conversely, beam splitters also can take separate beams and combine them into one beam. A prism uses the principle of light velocity in glass being every color, and therefore, wavelength dependent. This causes two or more colors entering a prism to all exit at different points. This color separation property works as an elementary color separator (See *Figure 8-17*) which uses two optoelectronic detectors, one for each color. An offset retroreflection prism is insensitive to rotation about an axis parallel to its roof. This type prism has a beveled roof so the beam is directed on center; however, a line across the beam will be missing. This prism uses its TIR or *total internal reflections*, as does a 90° turning beam prism. (See *Figure 8-18*) With the proper surface coatings you can realize losses of less than 1%.

Diffraction gratings are cuts or small surface grooves typically made on opaque materials. Each cut or grating purposely creates interference patterns as a light front strikes them. Therefore, this convention also works on a color or wavelength dependent relationship to spectrally separate colors. Some of the light striking the gratings passes through this opaque medium. Gratings however differ from prisms because gratings with multiple slits have places where their interference patterns cancel and other places where they reinforce one another. The result is some black and some bright spots respectively. Gratings further differ from prisms because they produces two light array patterns on each side of zero and the prism produces just one such pattern. (See *Figure 8-19*) *Binary optics* is the practice of purposely cutting small grooves in the normally smooth surface of a lens. The cuts look like staircases from the side, giving a desired image of staggered segments or other unusual effects, depending on the pattern cut.

Reflectors

Mirrors are obvious reflectors; however, curved reflectors such as *ellipsoidal* and *parabolic* also exist. An ellipsoidal reflector (See *Figure 8-20*) re-

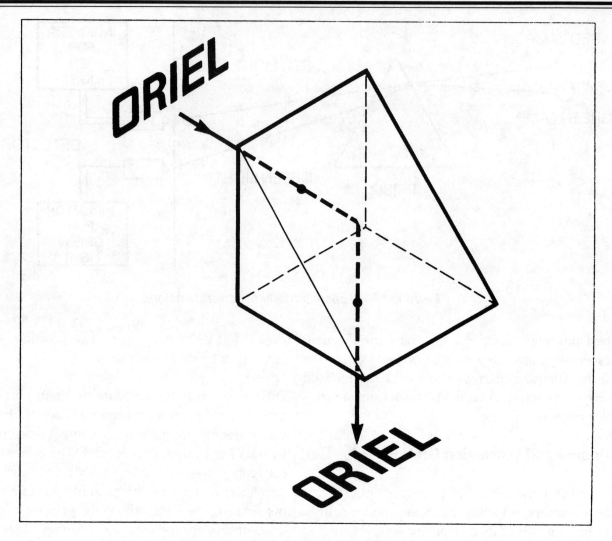

Figure 8-18. *A 90° light bending prism.*

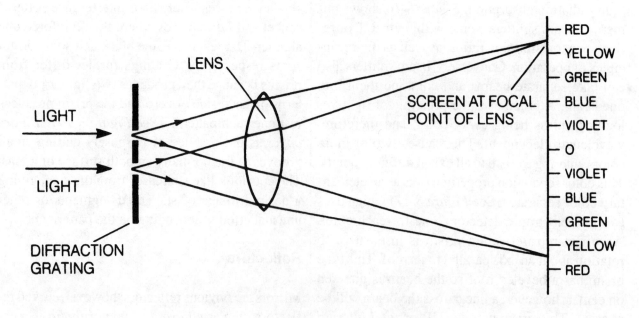

Figure 8-19. *A diffraction grating effect on colors.*

flects light from one focus to a second external focus. Note where the rays intersect near the reflector's back surface and where they converge at a point in front of the ellipsoidal reflector. A parabolic reflector collects radiation from a source at its focal point and reflects it as a collimated beam parallel to its axis.

Lens Surface Coatings

Earlier we assumed in our definition of a lens that it has virtually no reflections and totally passes light. However, to be absolutely precise, some reflections do exist, but coating the lens in its manufacturing process greatly reduces this and the glass's extremely small absorption of light. Lens coatings dictate three optical properties:

1. Transmittance.
2. Absorbance.
3. Polarization.

Coatings include metals, dielectrics and semiconductors. Evaporation is the process by which you actually apply these lens coatings. There are two kinds of evaporation:

1. Thermal vaporization.
2. Electron beam ion bombardment.

Thermal evaporation heats a material in a tungsten molybdenum "boat." The electron beam ion bombardment process bombards the coating material with an electron beam at up to 15 kV and several Amps current in a water cooled cooper crucible. A third technique sometimes used is sputtering which deposits thin films on usually larger lens areas.

The most important factor in this process is controlling the coating's thickness. You control this by interference monitoring and/or measuring the resonance of a crystal oscillator. Continuing evaporation beyond the 1/4 wave thickness increases transmittance to its original value up to a 1/2 wave thickness at which it momentarily stops. The crystal monitoring technique places a quartz crystal plate on the same plane as the coating material and the crystal's mass controls its resonant frequency. *Figure 8-21* shows an Oriel glass coated with 1/2 wave thickness peaking to the upper left of the reflectance curve and a 1/4 wavelength coating with slightly less than 50% of the 1/2 wavelength coating.

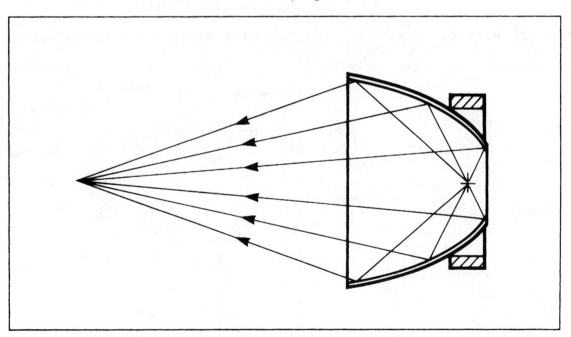

Figure 8-20. *An ellipsoidal reflector.*

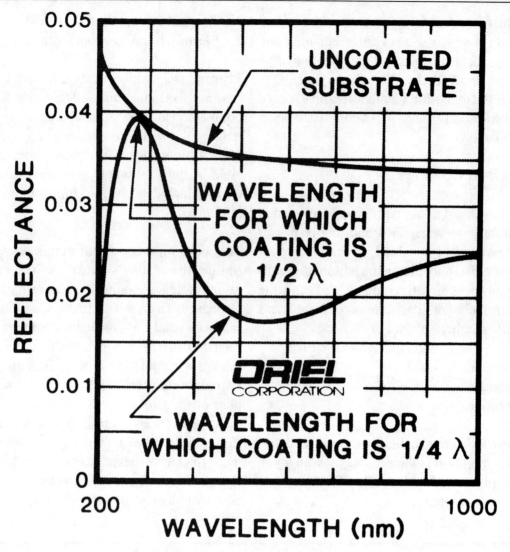

Figure 8-21. *A plot of a lens with a 1/4 and 1/2 wave thickness. (Courtesy of Oriel Corporation)*

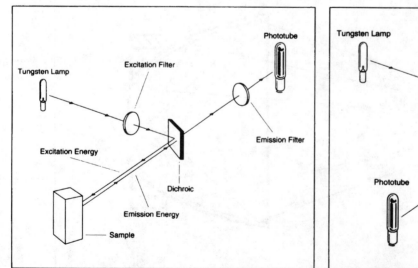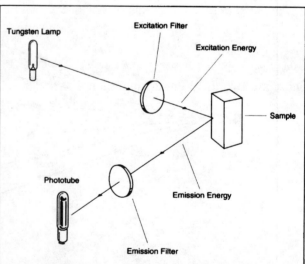

Figure 8-22. *Two different configurations of optical filters which purposefully block light.*

Coatings also exist on metal reflectors since it is nearly impossible to polish a metal's substrate to the extent required for specular (shiny) reflectance in the visible and UV regions. These specular or shiny surfaces are mainly concerned with the following properties:

1. Reflectance.
2. Durability.
3. Surface adherence.
4. Uniformity.
5. Long term stability.

Incidentally, all optical glass fluoresces under UV excitation. This is why UV sources predominantly use special fused silica or quartz glass lenses.

Filters With Blocking Glass

We have labored to pass as much light as possible through an optical glass lens; however, there are actually situations in which you do want to pass light through a glass element, specifically a filter. The majority of excitation and emission filters use one of two designs. In each case the function of the filter is to:

1. Allow transmission of energy at wavelengths within that of the filtering material.
2. Then reject it through either reflection or absorption within the emission spectrum which might later be mistakenly interpreted as emission energy. (See *Figure 8-22*)

There are two different arrangements of these filters. One uses a dichroic, or an optical element, which passes one wavelength and rejects another, and the other does not. These devices typically allow 50% or less light to pass through them.

F = Focal Length

$$\frac{1}{F} = \frac{1}{U} + \frac{1}{V} \qquad F = \frac{U \times M}{M + 1} \qquad F = \frac{U}{R + 1}$$

$$F = \frac{D \times R}{(R + 1)^2} \qquad F = \frac{U \times V}{D} \qquad F = \frac{D \times M}{(M + 1)^2}$$

D = Distance from Object to Image

$$D = F(M + \frac{1}{M} + 2) \qquad D = F(R + \frac{1}{R} + 2)$$

$$D = \frac{F(M + 1)^2}{M} \qquad D = \frac{F(R + 1)^2}{R}$$

U = Distance from Lens to Object

$$U = \frac{V}{M} \qquad U = \frac{D}{M + 1}$$

$$U = \frac{R \times D}{R + 1} \qquad U = \frac{F \times V}{V - F}$$

V = Distance from Lens to Image

$$V = U \times M \qquad V = \frac{F}{R} + F$$

$$V = \frac{U}{R} \qquad V = (F \times M) + F$$

M = Magnification

$$M = \frac{V}{U} \qquad M = \frac{F}{U - F} \qquad M = \frac{V - F}{F}$$

R = Reduction

$$R = \frac{U}{V} \qquad R = \frac{U - F}{F} \qquad R = \frac{F}{V - F}$$

Table 8-1. *The formulae describing simple single lenses. (Courtesy of Rolyn Corporation)*

Chapter 8 Quiz

1. Why don't mass produced optoelectronic devices use optical glass lenses?
   A. They can't withstand the temperatures.
   B. They are too expensive.
   C. A and B.
   D. None of the above.
2. A simple lens passes light virtually unimpeded but also can:
   A. Focus light.
   B. Reject all light.
   C. Both A and B.
   D. None of the above.
3. The EFL of a lens assumes:
   A. All rays of light are randomly scattered.
   B. All light reflects from its surface.
   C. All rays of light are parallel if the light source or object is far enough away from the lens.
   D. None of the above.
4. The M in the formulae in *Table 8-1* signifies:
   A. Multiplier effect.
   B. Multiple reflections.
   C. Magnification.
   D. None of the above.
5. The relative aperture number of a lens is its EFL divided by:
   A. The lens' clear aperture diameter.
   B. The lens' diopter number.
   C. The lens' transmittance.
   D. The lens' absorption factor.
6. Primary simple lens types depend upon the _____ of a lens(es).
   A. Spatial arrangement.
   B. Specular value.
   C. Grouping(s).
   D. All the above.
7. A Fresnel lens has:
   A. A series of concentric grooves.
   B. A short focal length.
   C. Minute refracting facets capable of bending light.
   D. All the above.
8. A material which blocks all light:
   A. Has a transmissivity of 0.
   B. Is opaque.
   C. Both A and B.
   D. None of the above.

9. Lens pairs predominately condense or _____ light.
   A. Deflect.
   B. Reflect.
   C. Focus.
   D. None of the above.
10. Outgassing is beneficial to mirrors, T or F?
11. A spherical mirror's focal length depends on its radius of curvature, T or F?
12. Prisms operate on the principle of different colors of entering light all having equal wavelengths, T or F?
13. Prisms work as elementary color separators, T or F?
14. One of the differences between a prism and diffraction grating is diffraction grating:
   A. Produces two light array patterns on each side of zero.
   B. Uses the interference principle to create black and bright spots.
   C. Both A and B.
   D. None of the above.
15. A car's headlights use which type of reflector? (You will have to analyze their properties to answer this question)
   A. Ellipsoidal.
   B. Parabolic.
   C. Ellipsoidal and parabolic in combination.
   D. None of the above.

Appendix A
An Optoelectronics Glossary

Appendix A
An Optoelectronics Glossary

Aberration — Rays from a zero-dimension object, such as a distant star, imaged through a perfect lens will all focus on a single zero-dimension spot. If the rays go anywhere else, that is an aberration.

Acceleration Factor — A factor which describes the change in a predicted phenomenon caused by a secondary effect.

Amorphous Silicon — A low cost form of silicon with a very disordered molecular structure, quite unlike the lattice molecular structure of silicon used in ICs and transistors. It is a core emerging solar technology.

Angle of Divergence — This is the spread of light after passing through a lens. The smaller the light source, the smaller the angle of divergence. A longer focal length reduces this angle.

Ångstrom — An Ångstrom (Å) measures length, especially electromagnetic wavelengths. One Å = 10^{-10} meters = 10^{-4} microns = $3.937 \cdot 10^{-9}$ inches.

ANSI — This is the American National Standards Institute, an industry-wide organization standardizing upon products, including most optoelectronic devices.

Area Source — A source with a diameter greater than 10% of the distance between it and the detector.

Avalanche Photodiode A photodiode operating on the principle of the avalanche multiplication photocurrent effect. It is very suited for applications requiring fast switching and/or low noise.

Backlighting — Illumination by an indicator of a front panel legend from behind, without the LED protruding from the front panel.

Band Gap — The potential energy difference between the conduction band and the valence band in a material. This determines the forward voltage drop and the frequency of light output of a diode.

Beam Angle — This is the angular spread of light from a lamp with angles ranging from 15° for a narrow spot to 130° for a floodlight.

Bicolor LED — A component that combines two dice of different colors upon a single substrate or lead-frame carrier. The device has either two or three lead wires for turning the device off and on.

Binary Optics — The practice of purposely cutting small grooves in the normally smooth surface of a lens. These cuts look like staircases from the side, giving a desired image in staggered segments or other unusual effects, depending on the pattern cut.

Blackbody — The device to which all irradiance measurements are referenced. A blackbody is a theoretically perfect radiator and absorber of radiant energy. Its radiation spectrum is therefore a simple function of its temperature.

Candela — A photometric unit of luminous intensity, expressed in lumens per steradian.

Candela/cm² — A luminance unit called a *stilb*.

Candle — This was the unit of luminous intensity for years and consisted of a wax candle about an inch in diameter. Today we better define this as one candle directed to a surface of one square foot, one foot away from the candle. It produces a uniform flow of light (flux) which is one lumen.

Candlepower — This is light intensity expressed in candles measured in one direction over a certain angle.

Chrominance Contrast — The color contrast between two adjacent surfaces of identical area, shape and texture. The human eye is more sensitive to difference in color than it is in brightness.

Collimated — The effect of concentrating a diffused light source into more of a beam shape.

Color Temperature — The temperature of a blackbody whose radiation has the same visible color as that of a given non-blackbody radiator, the measurement unit is K (Kelvin).

Concave Lens — A lens that is thinner in the middle than at its edges and tends to spread out light as it passes through the lens.

Concentrators — These emerging core solar technology devices focus sunlight on a small area (a thin strip of a silicon solar cell or a small "spot" solar cell). It converts sunlight to electricity without great quantities of silicon, lowering production costs.

Conduction Band — The empty energy band where electrons are the charge carriers.

Conservation of Energy Law — This states $E_R + E_A + E_T = 1$, or radiated plus absorbed plus transmitted energy must equal 1. An object with an emissivity of e = 0.80 has E_A = 80%; therefore $E_R + E_T$ = 0.20 or 20%.

Contrast — The noticeable difference in color, brightness, or other characteristics in a side-by-side comparison.

Contrast Ratio — The measurement of how visible, for example, a backlit legend is from its background. Its formula involves a ratio using both these quantities.

Convex Lens — A lens that is thicker in the middle than at its edges and tends to concentrate light as it passes through the lens.

Critical Angle — The maximum angle of incidence for which light will be transmitted from one medium to another. Light approaching the interface at angles greater than the critical angle will be reflected back into the first medium.

Crosstalk (Light Bleed) — The undesired illumination of one indicator position by an adjacent or different light source.

Crystalline Silicon — This traditional form of silicon has molecules positioned in a predetermined lattice. Crystalline structures yield higher efficiencies but cost more to produce and are soon to be abandoned in solar energy technology.

CTR (Current Transfer Ratio) — The ratio of the DC output current to the DC input current of an optically coupled isolator.

Dark Current — The leakage current of a photodetector with no radiation within its spectrum of sensitivity applied incident upon it. The smaller this leakage current, the better the optoelectronic semiconductor device.

Darlington Phototransistor — A pair of directly coupled transistors with the first transistor being a phototransistor. This device is very light sensitive.

Depletion Region — The area where the density of charge particles is negligible compared to the impurity concentration.

Detector Quantum Efficiency — The ratio expressing the number of carriers generated, divided by the number of photons absorbed.

Detector, Radiometric — A device which changes light energy (radiation) into electrical energy.

Die — The basic semiconductor device or "chip" inside the LED assembly.

Diffraction — The phenomenon of light bending at the edge of an obstacle, demonstrating wave properties of light.

Diffusant — Glass particles suspended in the epoxy lens of an LED which diffuse or broadly cast the light, increasing the viewing angle.

Diffuse Mode Photoelectric Sensors — These proximity detection photosensors are relatively inefficient, but are tolerant of the reflected beam's angle of arrival. They are very reflective sensitive, use no lens assemblies, and sense a white object at a much greater distance than a dark object, see Fixed-Field Photoelectric Sensors.

Diffusion — The clouded or scattered lens effect of an LED.

Diopter — A system which rates a lens by its refractive power instead of its focal length.

Dithering — The addition of a small signal, sometimes noise, to a system to improve performance. For example, many DVMs average a number of samples to give a more accurate measurement. Random noise helps, but a controlled signal provides better accuracy with fewer samples. This term also refers to modulation of light intensity.

Dobson Unit (DU) — A measure of how thick the ozone is in a column directly above the observer. A DU is the ozone thickness of 0.01 mm at $1°$ C and at 1 atmosphere. In America we experience about 300 DUs daily. It is named after the atmospheric scientist G.M.B. Dobson.

Dominant Wavelength — The wavelength that is a quantitative measurement of apparent light, as perceived by the human eye.

Doping — The addition of carrier supplying impurities to semiconductor crystals.

Dot Matrix — An array composed of addressable LED dots which can form numerous characters.

Duty Cycle — The measure of the effect of a pulsed input to a lamp, expressed as a percentage, of the ON time versus the total time.

Efficiency — The measure of the output power of a light source to its electrical input power.

Electroluminescence — The non-thermal conversion of electrical energy into light. In an LED, it is produced by electron-hole recombination in the P-N junction.

Electron Beam (EB) — This type of curing instantly cross-links adhesives, inks and coatings' polymers, more deeply penetrating a surface than mere UV curing alone.

Emissivity — The measure of how well an object absorbs or reflects radiant energy imparted upon it. Smooth shiny surfaces have emissivities approaching 0, which is a perfect heat reflector, while other objects have an emissivity value approaching 1.00, a perfect radiant energy absorber.

Emittance, Radiometric — Power radiated per unit area from a surface.

Epitaxial — Material added to a crystalline structure which has and maintains the original crystals' structure.

Erythema Action Curve — A plot, used in determining UV light and its subsequent ozone depletion's effect, of the light spectra of many biological actions.

Experimental UV Index — This Index, offered free by the National Weather Service and the EPA, is part of the weather report. It forecasts the amount of UV reaching the earth's surface in your locale at its peak hour, noon. In summer, it ranges from 0 to 15.

f Stop — This is the ratio of a lens' aperture (opening) to its focal length.

Fiber-Optics — These are optical fibers of transparent strands of glass or plastic used to conduct light energy into or out of hard to access areas. They also serve as a communications medium. A cladding material surrounds this glass or plastic, which is less dense than the cladding material and therefore has a lower index of refraction.

Fixed-Field Photoelectric Sensors — These photoelectric proximity sensors use a lens or lens assembly to purposefully limit their sensing range and compare the amount of reflected light seen by two different optodetectors. The angle of arrival of reflected light is very critical with this type photoelectric sensor.

Fixed Focus — This type photoelectric sensor detects an object in an area where emitting and receiving light cross.

Footcandles (fc) — An older irradiance (flux/area) term. A Lambert/meter^2, a more modern term, equals $9.29 \cdot 10^{-2}$ fc.

Foot-Lambert — A unit of luminance or brightness defined as the surface of one square foot upon which there is a uniformly distributed flux of one lumen, or lumen/ft^2.

Flashlamp — A device composed of glass bulbs and/or a metal can. It contains an anode, a cathode and trigger probes to guide the arc (flash). You may select the device's glass envelope and window to extend the light output into the ultraviolet range.

Flux — Power passing through a surface (energy per unit time). It is also the number of photons passing through a surface per unit time, expressed in lumens or watts.

Flux Density — The measure of a wave's strength. It is flux per unit area normal to the direction of flow, or the number of photons passing through a surface per unit area, expressed in watts/cm^2 or lumen/ft^2.

Fresnel — Augustin Fresnel (1788-1827) was a French scientist who invented a molded to shape lens, used as a light condenser, which now bears his name.

Fresnel Loss — As light passes from one medium to another, with a different index of refraction, the light lost which reflects back at this interface is Fresnel loss.

Gate — The control terminal of an SCR, or a logic function component.

H — This is an irradiance or radiation flux density expressed in watts/cm^2.

Hole Electron Pair — A positive (hole) and a negative (electron) charge carrier, considered together as an entity.

Incandescence — Emission of light by thermal excitation resulting from the superheating of a conductor. The excitation must be sufficient to produce enough photons to make the light visible.

Index of Refraction — The ratio of the speed of light in a vacuum to the speed of light through another material, e.g. a plastic lens.

Illumination — Light level on a unit area.

Infrared Radiation — The electromagnetic wavelength region between approximately 0.75 and 100 micrometers which is longer than visible light.

Injection Laser Diode — A P-N semiconductor device which uses lasing to increase the light output and concentrate the light in a small area.

Intensity — The radiant flux emitted by a light source per unit solid angle, i.e. lumens or foot-candles per steradian.

Interrupter Module — An electronic device producing an electrical output when an object breaks the path of a beam of visible or non-visible light between a photo source and detector. The emitter/detector pair is usually housed in a module with a slot.

Inverse Square Law — The illumination on a surface varies inversely as the square of the distance. As an example, a light source three feet away is only 1/9 as bright as the source at one foot.

IR LED — An infrared LED consisting of a semiconductor P-N junction emitting light, when forward biased, in the range of 0.78 mm to 100 mm.

Irradiance — The radiant flux density incident upon a surface. The ratio of flux to area of an irradiated surface, typical units are: W/ft^2 or W/m^2, with $1 \ W/ft^2 = 10.764 \ W/m^2$.

Isolation Voltage — The dielectric withstanding voltage capability of an optocoupler under defined conditions and time.

Joules — Energy is measured in Joules. The rate at which work is done is energy per unit time. A work rate of one Joule per second is defined as one watt. Many optical instruments collect all the light incident upon them and divides this by the detector area in cm^2, then integrate or average this over the exposure time in seconds which equals energy per unit area ($Joules/cm^2$).

Kelvin — A temperature scale starting at absolute zero where no molecular movement occurs. Zero Kelvin (no degree sign is used) is -273.15° C and each Kelvin unit equals 1° C; therefore, 0° would be 273.15 Kelvin.

(1/p) candela/cm² — A unit of luminance called a Lambert.

Lambert's Law or the Law of Cosines — This describes the spatial relationship of a perfectly diffusing surface upon which light may be either emitting or reflecting. The light emitted or accepted by the surface decreases with the cosine of the angle from a perpendicular to the surface.

Lasing — Stimulating early recognition of carriers to emit radiation in the same direction as, and coherent with, some initial stimulating radiation. It represents amplification which preserves the direction, frequency and phase of the amplified light.

LASER (Laser) — This acronym for Light Amplification by Stimulated Emission of Radiation describes a device which self stimulates and produces a very pure red light when it experiences a sufficient electric current.

Laser Efficiency — A method of evaluating the efficiency of a laser diode. Five methods are used as follows:
1. Internal Quantum Efficiency: Photons generated per current carrier injected.
2. Emission Efficiency: Photons emitted per photon emitted.
3. External Quantum Efficiency: Photons emitted per carrier current injected. This is also called "junction efficiency."
4. Differential Efficiency: The slope of the light output versus the forward current curve above the lasing threshold. Also called "slope efficiency" or "incremental efficiency."

5. Power Efficiency: Light output power divide by total input power. Also referred to as "device efficiency," "overall efficiency," or "conversion efficiency."

Lens — A curved piece of transparent material, typically glass or plastic, which bends rays of light passing through. Or, the epoxy molded to an LED die to provide a certain desired optical characteristic.

Light Current — This is the current flowing through a photosensitive semiconductor, such as a photodiode or phototransistor when exposed to illumination or irradiance.

Light Curtain — This is light produced by a line or predetermined array of emitting and receiving photoelectric sensors which detect objects within their sensing field.

Light Emitting Diode (LED) — A semiconductor diode emitting incoherent light at its P-N junction when forward biased.

Light Pipe — An optical conduit made of molded plastic that directs light from an LED to the desired viewing location, often this is at a right angle from the LED's circuit board.

Liquid-Phase Epitaxy — A process by which epitaxial are grown on substrates while immersed in liquid gallium at high temperatures.

Lumen — The unit of luminous flux measuring the flow or quantity of light. One lumen describes the light on a surface one foot square located one foot away from a one candle source.

Luminance Contrast — The observed brightness of a light emitting element compared to the brightness of the surroundings of the device.

Luminescence — The emission of light due to any cause other than temperature (heating), which is incandescence.

Luminous Flux — The time rate of flow of light. The CIE curve relates luminous flux to radiant flux, as perceived by the eye's response curve.

Luminous Intensity — Luminous flux per unit solid angle in a given direction.

Majority Carriers — Charge carriers responsible for conduction under thermal equilibrium; electrons in N-type or holes in P-type materials.

Mark Sensors — These photoelectric light sensors use an optical lamp to detect colors by distinguishing between the sensed object and its background color.

Miller Effect — The phenomenon of decreased switching speed due to inter-electrode capacitance in an optoelectronic device. Its magnitude is this capacitance times the device's gain.

Minority Carriers — Electrons in P-type material or holes in N-type material.

Mobility — The velocity of a charge carrier per unit of applied electric field.

Modulated LEDs — LEDs can turn off and on at a far greater frequency than possible with incandescent lamps. This allows the amplifier of a phototransistor receiver to "tune" to this modulated LED frequency and amplify only light signals pulsing at that frequency.

Modulation — The transmission of information by modifying a carrier signal, usually its amplitude or frequency (AM and FM).

Monochrome — Any combination of colors of the same hue, but of different saturations and luminances.

Monochrometer — An instrument which is a source of any specific wavelength of radiation over a specific band.

Monochromatic — Of a single color or wavelength.

Nanometer — A unit of length (10^{-9} meters) used as a unit of wavelength of light. It is related to the color perceived by the eye. A nanometer is equal to 10 Ångstroms.

Narrow View Reflective — This type of photoelectric sensor has a very narrow sensing field and is used for more precise detection than possible with an ordinary diffuse reflective light sensor.

Neon Lamp — A light source that generates blue or amber light by thermally exciting a neon gas plasma with heated electrodes.

Noise Equivalent Power (NEP) — This is a means of expressing an optoelectronic device's spectral response by equating it to the light level required to obtain a S/N ratio of one.

Numerical Aperture — The sine of half the angle of light acceptance.

Opposed Mode of Photoelectric Sensing — This mode of sensing has the photo-emitter and photo-detector placed directly opposite each other and an object is detected when interrupting this beam of light.

Optocoupler/Opto-isolator — An optoelectronic semiconductor device transmitting electrical signals without an electrical connection between the light source (input) and a light detector (output). The input is generally an LED. The output may assume a variety of different type devices such as a photodiode, phototransistor, photodarlington pair transistor, etc.

Peak Spectral Emission — The wavelength of highest intensity of a light source.

Photoconductor — A material with resistivity which varies with changing illumination levels.

Photo Current — The difference between light current and dark current in a photodetector.

Photodarlington — A light-sensitive transistor pair connected with very high light sensitivity to illumination and radiation.

Photodiode — A semiconductor device which conducts when forward biased and when incident light falls upon its surface. The relationship between its input radiation and output current is very linear.

Photodetector — An optoelectronic device producing an electrical signal when subjected to radiation in the visible, infrared or ultraviolet regions. Photodiodes, phototransistors, and photodarlington pairs are examples of this type device.

Photometer — An instrument which measures <u>visible</u> light brightness (intensity), usually in Lumens/cm^2, and the quantity (flux) of light.

Photometry — The measurement of visible light in quantity (flux) and brightness (intensity).

Photomultiplier Tube — A photoemissive type photosensor, encased in a vacuum tube, which emits one electron per each photon falling upon a metal photo cathode. It uses successive amplification stages, using secondary emission, to amplify otherwise minute electron current.

Phonon — In the absorption process, electrons move from the valance to the conduction band, giving up heat (phonons) and light (photons). A phonon is only about 0.05 eV, 1/20th that of a photon.

Photon — A quantity (the smallest possible unit) of radiant energy. A photon carries a quantity of energy equal to Planck's Constant times frequency.

Photothyristor — A thyristor whose switching action is controlled by light applied to the thyristor's gate.

Phototransistor — A light-sensitive transistor producing an electrical signal in proportion to the intensity of the applied light. This low level photocurrent is amplified by the current gain of the transistor (or gate if the transistor is a FET). The base, or gate if it is a FET, may or may not be brought out of the case for control purposes.

Photovoltaic — A type of photosensor which generates a voltage across a P-N junction as a result of photons falling upon this P-N junction. A solar cell is a photovoltaic sensor (generator).

Photovoltaic Cell — A photosensitive device which supplies DC electricity when illuminated by radiant energy.

Photovoltaic Effect — The generation of voltage from incident radiant energy (typically the sun) from the use of dissimilar materials, one of which is light-sensitive material.

Planck's Law — This law states you can plot radiated energy as a direct function of wavelength.

Plasma — This is merely ionized gas.

Point Source — A radiation source whose maximum dimension is less than 1/10 the distance between the light source and detector.

Polarizing Filters — These antiglare filters emit only vertically polarized light. Reflected light comes back rotated by 90° so it only accepts horizontally polarized reflected light.

Prism — A device which separates light into its spectral components. In LEDs, the prism directs light output from an LED to the viewing location.

Proximity Mode of Photoelectric Sensing — This mode senses an object directly in front of it by detecting the sensor's own energy reflected back to the optodetector. This mode establishes a light beam, rather than detecting a broken beam.

Quantum Efficiency — The ratio of the number of carriers generated to the number of photons incident upon the active region.

Quantum Theory — The concept that light is emitted in minute bundles of energy rather than doled out in a steady stream.

Radiant Flux — The time rate of flow of radiant energy.

Radiometer — An instrument which measures *non-visible* light brightness (intensity), usually in Lumens/cm², and the quantity (flux) of light.

Radiometry — The measurement of non-visible light.

Recombination — The combining of a hole and an electron.

Reflector Module — An electronic device containing a light source and a photodetector which detects any object which reflects light back to the detector.

Refractance — The phenomenon of light bending as it passes through one medium to another, such as air to water.

Resolution — The number of visible distinguishable units in the optoelectronic device's coordinate space.

Retroreflective Mode of Photoelectric Sensing — This type photoelectric sensor contains both the photoemitter and the photodetector in the same housing. The detectors senses a reflected beam off the surface and is used to detect an object moving past, breaking this reflected beam.

Secondary Optics — Devices used to enhance or redirect an LED's light output. Examples include lenses and light pipes.

Silicon Film — This silicon, grown in long planks, is an emerging core solar technology which uses thinner silicon layers to cut costs of manufacturing.

Source — A light source which provides radiant energy.

Spatial — The directional characteristic of light in space.

Spectral Distribution — The distribution of light by wavelength with an electromagnetic spectrum.

Spectral Output — This describes the radiant energy or light emission characteristic versus the wavelength of a device.

Spectral Sensitivity — A plot of the light detector's sensitivity versus the wavelength detected.

Specular — This describes a highly reflective smooth shiny surface.

Stefan-Boltzmann Equation — Heating mass releases detectable heat, and the Stefan-Boltzmann equation describes the amount of this energy released.

Steradian — A solid angle of a sphere encompassing a surface area equal to the square of the radius of the sphere. There are 12.56 steradians on a sphere.

Temporal — The characteristic of light with time.

Thermal Detectors — Thermocouples and thermopiles are examples and these devices make radiometric measurements through the incident radiation which heats them.

Thermopile — A very broadband, heat sensing, radiation detector.

Tint — A color added to an LED's epoxy lens to identify it when it is ON.

Total Flux — This is the flux emitted in all directions. Candlepower ratings refer to flux in one specific direction.

Trapping — The capturing of a hole or an electron in an impurity or defect.

Trigonometric Reflective — These photoelectric light sensors emit light which hits the sensed object and then measure the transmitted and reflected light's difference angle.

Triple-Cell or Triple-Junction — These devices are layered three deep to purposely trap shorter wavelengths of light which were otherwise escaping. This new emerging core technology is intent on enhancing solar energy conversion efficiency.

Tungsten — The element normally used for incandescent lamp filaments. Light standards use special calibrated tungsten bulbs.

UV — This is the sun's radiant energy and occurs in three classes or radiation bands:
1. UV-A is between 320 and 400 nm.
2. UV-B is between 280 and 320 nm.
3. UV-C is between 200 and 280 nm. The shorter the wavelength, the more destructive or biologically damaging the UV.

VUV (Vacuum UV) — This UV is man-made in a vacuum and has a lower wavelength, higher frequency, and is proportionately more energetic with a far greater potential for biological damage. It is used to scribe (etch) and clean semiconductor substrates.

Valence Band — The filled energy band from which electrons are excited into the conduction band.

Wafer — A semiconductor substrate with an epitaxial layer on it.

Water Clear LED — An LED die combined with a clear lens that has no tinting.

Wavelength — The velocity of a wave divided by its frequency.

Wein's Displacement Law — This equation inversely relates wavelength to an object's temperature, expressed in Kelvin. This law explain the wavelength of peak energy shifting to the shorter wavelength end of the scale.

Appendix B
Sources of Supply

Appendix B
Sources of Supply

EXAMPLES/MANUFACTURERS OF SOLDER FLUXES/CONDITIONERS

Ultra Mild-Type R: Alpha 100
Minor-Type RMA: Alpha 611 or Kester 197
Mild-Type RA: Alpha 711-35, Alpha 809 foam, Kester 1544 or 1585
Moderate-Type AC: Alpha 830 or 842, Kester 1429/1429 foam, Lonco 3355

SURFACE CONDITIONERS

Alpha 140 or 174, Kester 5560, Lonco TL-1

ADDRESSES OF MANUFACTURERS OF LED TREATMENT RELATED PRODUCTS

Alpha Metals, Inc.
56 G Water Street
Jersey City, NJ 07304
302-434-6778

London Chemical Co. (Lonco)
240 G Foster
Bensenville, Illinois 60106
312-287-9477

The Orchard Corporation
(Silver Saver)
1154 Reco Avenue
St. Louis, Missouri 63126

Kester Solder Co.
4201 G Wrightwood Ave.
Chicago, Illinois 60639
312-235-1600

Allied Chemical Corporation
Specialty Chemicals Division
P. O. Box 1087R
Morristown, NJ 07960
201-455-5083

MANUFACTURERS OF PHOTOCELLS AND PHOTODIODES

EG&G Vactec Optoelectronics
10900 Page Blvd.
St. Louis, MO 63132
314-423-4900

Centronic Inc.
2088 Anchor Court
Newbury Park, CA 91320
805-499-5902

ADDRESSES OF IR PYROMETER MANUFACTURERS

Watlow
12001 Lackland Road
St. Louis, MO 63146
314-878-4600

Raytek
1201 Shaffer Road
P.O. Box 1820
Santa Cruz, CA 95061-1820
800-227-8074

Wahl Instruments, Inc.
5750 Hannum Avenue
Culver City, CA 90230
800-421-2853

Gentri Controls, Inc.
19 Ben's Way
Hopedale, MA 01747
508-634-3511

MANUFACTURERS' ADDRESSES OF IRDA RELATED PRODUCTS

ACTiSYS Corp.
1507 Fulton Place
Freemont, CA 94539
510-490-8024

Genoa Technology
5401 Tech Circle
Moorpark, CA 93021
805-531-9030

Hewlett-Packard
Corvallis, Oregon
800-677-7001

Irvine Sensors Corp.
3001 Redhill Ave.
Building III
Costa Mesa, CA 92626
714-549-8211

Puma Technologies
3375 Scott Blvd. Suite 300
Santa Clara, CA 95054
408-987-0200

TEMIC
A company of Diambler-Benz
U.S. Representative:
Siliconix
2201 Laurelwood Road
P.O. Box 54951
Santa Clara, CA 95056
408-988-8000

SOURCE OF SUPPLY FOR IR SENSOR CARDS

Laser 2000 GmbH
Argelsrieder Feld 14
82234 Wessling Germany
Tel +49 8153/405-0

Siemens Components
Optoelectronics Division
19000 Homestead Road
Cupertino, CA 95014

The UV Experimental Index

EPA Stratospheric Ozone Hotline:
1-800-296-1996

The National Weather Service:
1-301-713-0622

SOURCE OF SUPPLY FOR NIGHT GOGGLE FILTERS

WAMCO Inc.
11555-A Coley River Circle
Fountain Valey, CA 92708
714-545-5560

SOURCES OF SUPPLY FOR UV-BASED PRODUCTS

UV Xenon Flashlamps

EG&G Electro-Optics
35 Congress Street
Salem, MA 01970
508-745-3200

Hamamatsu Corp.
360 Foothill Road
P.O. Box 6910
Bridgewater, N.J. 08807-0910
1-800-524-0504

Credit Card Sized Sensometer

Mr. Rick Giese
P.O. Box 540
Kula, Hawaii 96790
SouthSky@Maui.net
1-800-96-HAWAII

UV Intensity-Light Controller

Oriel Instruments
250 Long Island Blvd.
P.O. Box 872
Stratford, CT 06497
203-377-8282
res sales@oriel.com

UV Analytical Software

Sensor Physics Inc.
attn: Gary Forrest
105 Kelleys Trail
Oldsmar, FL 34677
813-781-4240
103154.266@compuserve.com

Diesel Engine and Radiator Leak Detection by UV

UVP, Inc.
2066 W. 11th Street
Upland, CA 91786
800-452-3597

UV-Sensitive CCD Technology

PixelVision, Inc.
attn: George M. Williams
15250 NW Greenbrier Parkway
Suite 1250 Beaverton, OR 97006
503-629-3210

UV-BASED INSTRUMENTS

UV-Dosimeter

The LightBug™
International Light
17 Graf Road
Newburyport, Mass. 01950
508-465-5923

Miltec Corp.
303 Najoles Rd. Suite 108
Millersville, Md. 21108
800-999-2700

EIT Instrumentation Products
108 Carpenter Drive
Sterling, Va. 20164
703-478-0700
www.eitinc.com

UV-B Biological Effect Meter

Solar Light Company
721 Oak Lane
Philadelphia, PA 19126-3342
215-927-4206
CompuServe: 72073,2737

UV-Based Curing

Fusion UV Curing
AETEK Manufacturing, Inc.
7600 Standish Place
Rockville, Md. 20855-2798
301-251-0300

Source of UV-Based Papers

Society of Manufacturing Engineering
One SME Drive
P. O. Box 930
Dearborn, Mich. 48121
313-271-1500

Radtech International
North America
60 Revere Drive Suite 500
Northbrook, Ill. 60062
847-480-9080

UV-Based Flame Detectors

Spectrex, Inc.
Peckman Industrial Park
218 Little Falls Road
Cedar Grove, N.J. 07009
website: http://www.spectrex-inc.com

Detector Electronics Corp.
6901 West 110th Street
Minneapolis, MN 55438
1-800-765-FIRE

Sierra Safety Technology
702-267-2960
webmaster@flametech.com

Spectral Sciences, Inc.
99 South Bedford Street
Burlington, MA 01803
617-273-4770
sag@spectral.com

UV Curable Adhesives

Blaze Technology Pte Ltd.
65A Jalan Tenteram #07-90
Saint Michael Ind. Estate
Singapore 328958
(65)-252-3568
email: BLAZE Technology

MANUFACTURERS OF OPTOELECTRONIC DEVICES

EG&G Sensors
Miltipas, CA 95035-7416
800-767-1888
Santa Clara: 408-988-8000
Northbrook: 847-480-9080
Cupertino: 408-257-7910
Hackensack: 201-489-8989
Palo Alto: 415-857-1501
San Diego: 619-549-6900

Scientific Technologies Inc.
31069 Genstar Road
Hayward, CA 94544-7831
800-221-7060

Siemens Components
Optoelectronics Division
19000 Homestead Road
Cupertino, CA 95014
408-257-7910

Industrial Devices, Inc.
260 Railroad Avenue
Hackensack, N.J. 07601
201-489-8989

Martech
120 Broadway
Menand, N.Y. 12204
1-800-362-9754

Texas Instruments
P. O. Box 660199
Dallas, TX 75266-0199
214-917-1264

Motorola
Literature Division
P. O. Box 20912
Phoenix, AZ 85036
1-800-441-2447
http://Design-NET.com

Hewlett-Packard
P. O. Box 10301
Palo Alto, CA 94303-0890
415-857-1501

Photocell Manufacturer

Clairex Electronics
560 South Third Street
Mount Vernon, NY 10550
914-664-6602

Optoelectronic Sensing

SUNX
1207 Maple
West Des Moines, Iowa 50265
1-800-280-6933

LED SECURING DEVICES AND LENS CAPS

Dialight
1913 Atlantic Avenue
Manasquan, NJ 08736
908-223-9400

APM Hexseal
44 Honeck Street
Engelwood, NJ 07631
201-569-5700

Bivar, Inc.
4 Thomas Street
Irvine, CA 92618
714-951-8808
bivar@interserv.com

Visual Communications Co. Inc.
7920-G Arjons Drive
San Diego, CA 92126
619-549-6900

OPTOELECTRONIC MEASURING INSTRUMENTS

Tektronix
P.O. Box 500
Beaverton, Oregon 97077
1-800-872-7924 or
1-800-547-5000

International Light
17 Graf Road
Newburyport, Mass. 01950
1-508-465-5923

PHOTODIODE AND PHOTODETECTOR MANUFACTURERS

Advanced Photonix, Inc.
1240 Avenida Acaso
Camarill, CA 93012
805-484-2884

Burle Industries, Inc.
1000 New Holland Ave.
Lancaster, PA 17601-5688
717-295-6771

Centronic, Inc.
2088 Anchor Court
Newbury Park, N.J. 91320-1601
805-499-5902

Detection Technology, Inc.
Valkjarventie 1
FIN-02130 Espoo
Finland
358 0 455 5600
info@dti.fi

Electron Tubes
100 Forge Way, Unit 5
Rockaway, N.J. 07866
201-575-5586

Hamamatsu Corp.
360 Foothill Road
P.O. Box 6910
Bridgewater, N.J. 08807-0910
1-800-524-0504

SOURCE FOR MOST OF THE PROJECTS' COMPONENTS

Digi-Key
701 Brooks Ave. South
P. O. Box 677
Thief River Falls, MN. 56701-0677
800-344-4539

SOURCES OF LENSES AND OPTICAL COMPONENTS

Sources of Optics Tutorials

Oriel Corp.
250 Long Beach Blvd.
P.O. Box 872
Stratford, Ct. 06497
203-377-8282

Rolyn Corp.
706 Arrowgrand Circle
Covina, CA 91722
818-915-5707

Optical Components Tutorial
Newport Corporation
1791 Deere Avenue
Irvine, CA 92606
800-222-6440
www.newport.com/tutorials/note2.html

A Convex Lenses Tutorial
Prof. Selman Hershfield
University of Florida
selman hershfield/selman@phys.ufl.edu
http://cpcug.org/user/laurence/sciteach.html

KEY SOLAR RESEARCH COMPANIES AND THEIR AREAS OF EXPERTISE

AstroPower, Inc.
Newark, Delaware
Developed silicon-film cells
Phone: 302-366-0400.

Energy Conversion Devices
Troy, Michigan
Working on roll-to-roll amorphous silicon
manufacturing and triple-cell structures
Phone: 313-280-1900 or 313-362-4780.

Entech, Inc.
Dallas, Texas
Designed a new line-focus concentrator which
uses 3M Linear Lensfilm to concentrate sunlight
Phone: 817-481-5588.

Siemens Solar Industries
Camarillo, California
Working on techniques to reduce the "kerf" loss
or waste in sawing silicon wafers, and also in
developing thin wafers
Phone: 805-482-6800.

Solarex Thin Film Division
Newtown, Pennsylvania
Developing a triple-junction thin film using less
material, and realizing an enhanced efficiency
Phone: 215-860-0902

ADDRESSES OF WRIST INSTRUMENT MANUFACTURERS

Breitling U.S.A.
1-203-327-1411

Casio, Inc.
570 Mount Pleasant Ave.
Dover, N.J. 07801
201-361-5400
1-800-634-1895

Cygnus, Inc.
400 Penobscot Drive
Redwood City, CA 94063
415-369-4300
415-599-3565
FAX 415-599-2503

Polar Electro Inc.
2501 West Burbank Blvd. #301
Burbank, CA 91505
818-563-2865
FAX 818-563-2867

Seiko Communications of America
1625 NW Amber Glen Court
Ste 140
Beaverton, Oregon 97006
503-531-1623
FAX 503-531-1550

Swatch TelecomA Division of SMH (US) Inc.
35 East 21st Street
New York, New York 10010
1-800-8-SWATCH

Timex Watch Co.
Park Extension Road
Post Office Box 310
Middlebury, CT 06762
203-573-5764
FAX 203-573-4883

Appendix C
Answers to All Quizzes

Appendix C
Answers to All Quizzes

Chapter 1 Quiz

1. T
2. T
3. F
4. F
5. T
6. D
7. A
8. D
9. C
10. B
11. B
12. A
13. A
14. A
15. D
16. B
17. A
18. A
19. C
20. A

Chapter 2 Quiz

1. T
2. F
3. T
4. T
5. T
6. B
7. C
8. T
9. F
10. A
11. T
12. F
13. T

14. C
15. T
16. C
17. C
18. C
19. A
20. B

Chapter 3 Quiz

1. F
2. T
3. T
4. F (cosine)
5. T
6. C
7. D
8. B
9. T
10. T
11. F
12. F
13. D
14. A
15. D

Chapter 4 Quiz

1. D
2. A
3. C
4. B
5. D
6. B
7. C
8. A
9. C
10. A

11. B
12. A
13. C
14. B
15. A
16. D
17. B
18. B
19. C
20. D
21. C
22. C
23. D
24. A
25. A
26. D
27. A
28. B
29. T
30. T
31. F
32. T
33. F
34. F
35. F
36. F
37. F
38. D
39. C
40. T
41. T
42. C
43. B
44. F
45. F
46. F
47. F
48. F

49. T
50. F
51. D
52. B
53. F
54. F
55. F
56. T
57. F
58. T
59. T
60. F
61. C
62. D
63. A
64. C
65. D
66. B
67. A
68. D

Chapter 5 Quiz

1. B
2. D
3. B
4. C
5. T
6. A
7. A
8. D
9. F
10. T
11. B
12. T
13. C
14. B
15. A
16. C
17. D
18. A

Chapter 6 Quiz

1. B
2. D

3. C
4. A
5. A
6. B
7. B
8. D
9. C
10. B
11. C
12. B
13. A
14. D
15. D

Chapter 7 Quiz

1. B
2. C
3. D
4. B
5. A
6. T
7. T
8. A
9. C
10. C

Chapter 8 Quiz

1. B
2. A
3. C
4. C
5. A
6. C
7. D
8. C
9. C
10. F
11. T
12. F
13. T
14. C
15. B

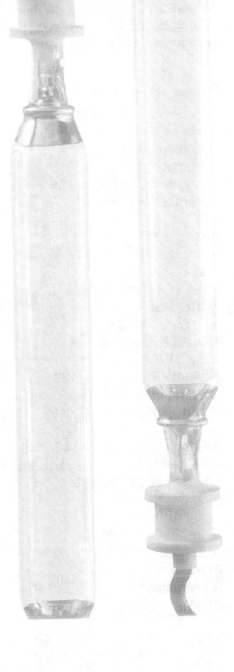

Index

Index

TN 194
Toggling 100
Tolerance 107, 148
Total Internal Reflections 215
Trace 57, 73
Transients 77
Transistor 24, 34, 35, 69, 81, 88, 98, 104, 117, 126
Transistor-Based Astable Oscillator 84
Transit Time 160
Transition 27, 32
Transmissivity 52, 210
Transmittance 204, 209, 217
Transmitter 27
Transparent Substrates 34
"Trapping" Technique 139
Tri-State™ RAMs 120
Triads 172
Triangular Wave 83
Triangular Wave Oscillator 110
Tungsten Deposits 78
Tungsten Filament Lamp 149
TV 186
Twisted Nematics 194

U

Ultraviolet 9, 26, 32, 57
Ultraviolet Detector 157
Ultraviolet Filter 190
Ultraviolet Light 26
Uniformity 219
Unit Symbol 8, 10, 19, 50, 149
Unit Symbols, Prefix 10
Unitary Luminance Difference 187
Unity Gain 82
Utility Grid 142
Utility Impedance 142
Utility Voltage 141
UV 9, 10
UV-C Region 157
UVtron 156

V

Vactec 147, 149
Vacuum 138
Vacuum Glass Envelope 77
Vacuum Tubes 61
Valence 25, 27, 30, 31, 35, 38
Valence Band 33
Valence Electrons 26

Valence Shell 25
Vapor 36
Vapor Deposition 34
Vapor Phase Epitaxy 40
Variable Duty Cycle 69
Variable Duty Cycle Oscillator 88
Vector Displays 176
Vector Storage 172, 173
Vectors 172, 174, 176
Vertical Axis 196
Vertical Gain 57
Vertical Input 57
Video Amplifiers 175, 176
Viewing Angle 18, 99, 187, 189, 194
Virtual Ground 106
Virtual Ground Potential 162
Visible LED 46, 50
Visible Light 8, 9, 10, 26, 33, 46, 69, 99, 138, 157, 161
Visible Light Emission 24
Visible Light Output 69
Visible Light Sources 68
Visible Spectrum 48
Visible Wavelengths 9
VLED-1 62
VLS 68, 69, 99, 113, 114, 116, 123
Volatile Memory 119
Volt 8
Voltage 90, 92, 101, 105, 108, 109, 114, 145, 177
Voltage Capability 140
Voltage Divider 107
Voltage Divider Resistors 107
Voltage Reference 107
Volume Units 108
VPE 40

W

Wafer Fabrication 37
Wafers 34, 36, 37, 139
WAMCO 193
Water Clear LEDs 69
Water Pumps 142
Watt 8
Wave 8, 9
Wave Shape Oscillator 110
Wave Shaping Circuits 68
Wavelength 8, 10, 26, 27, 32, 33, 34, 35, 39, 48, 53, 56, 61, 144, 147, 158, 159, 184, 185, 193

The Microcontroller Beginner's Handbook
Lawrence A. Duarte

Microcontrollers are found everywhere — microwaves, coffee makers, telephones, cars, toys, TVs, washers and dryers. This book will bring information to the reader on how to understand, repair, or design a device incorporating a microcontroller. *The Microcontroller Beginner's Handbook* examines many important elements of microcontroller use, including such industrial considerations as price vs. performance and firmware. A wide variety of third-party development tools is also covered, both hardware and software, with emphasis placed on new project design. This book not only teaches readers with a basic knowledge of electronics how to design microcontroller projects, it greatly enhances the reader's ability to repair such devices. Lawrence A. Duarte is an electrical engineer for Display Devices, Inc. In this capacity, and as a consultant for other companies in the Denver area, he designs microcontroller applications.

Electronic Theory
240 pages ▲ Paperback ▲ 7-3/8 x 9-1/4"
ISBN: 0-7906-1083-3 ▲ Sams: 61083
$18.95 ($25.95 Canada) ▲ July 1996

COMING SPRING 1997!
Optoelectronics
Volume 2
Vaughn D. Martin

This book is a comprehensive guide to intermediate electro-optics. It moves the reader into the more advanced topics of light and the electromagnetic spectrum. Also discussed are the concepts of reflection, refraction, interference, polarization, and diffraction.

One of the most important features of this book, however, is the detailed instructions it provides for integrated photodiodes and practical circuits. Many examples are given that use both transmitter and receiver circuits, and various appendices contain information on the operating characteristics of op-amps, light sources, and photodiodes. *Optoelectronics* is intended for students, electrical and mechanical engineers, managers, and technicians who need a basic understanding of optoelectronics and the techniques required to deal with common problems and errors.

Electronic Theory
400 pages ▲ Paperback ▲ 8-1/2 x 11"
ISBN: PENDING ▲ Sams: PENDING
$29.95 ▲ March 1997

Power Supplies
David Lines

Power supplies, the basic sources of energy in all electronic equipment, are essential to electronic design and construction. *Power Supplies* will guide the reader from the fundamentals of power supply components and their functions to the design and construction of a power supply system. The comprehensive coverage includes the basics of AC and DC energy sources, regulated and unregulated power supply systems, and linear and switching power supply projects. *Power Supplies* has two goals. First, to help the reader understand the basic function of each of the components in a power supply. And secondly, to show readers how to build useful working power supplies.

David Lines has been involved in various aspects of the electronics industry as a designer and technical writer, and as the owner of Lines' Logic Lines, a consulting firm. He has participated in the microprocessor-control revolution, documenting computer-controlled mass-transit systems.

Electronic Theory
92 pages ▲ Paperback ▲ 6 x 9"
ISBN: 0-7906-1024-8 ▲ Sams: 61024
$10.95 ($14.95 Canada) ▲ December 1992

CALL 1-800-428-7267 TODAY FOR THE NAME OF
YOUR NEAREST PROMPT PUBLICATIONS DISTRIBUTOR

▼   ▼   ▼   ▼   ▼   ▼   ▼   ▼   ▼   ▼   ▼

Simplifying Power Supply Technology
Rajesh J. Shah

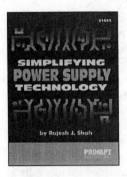

Power supply technology has come a long way in recent years. Many researchers have dedicated their careers to advancing the field of power electronics, and a good amount of effort is put into making contributions to the field. One can easily appreciate this field by understanding the basic concepts introduced in this book. *Simplifying Power Supply Technology* is an entry point into the field of power supplies. It simplifies the concepts of power supply technology and gives the reader the background and knowledge to confidently enter the power supply field. This book is also useful to experienced technicians as a reference to the basic concepts of various topologies.

Rajesh J. Shah is currently working for AC Delco Systems on the General Motors electric car program. He has twelve years of experience in the field of power conversions, and holds a master's degree in Electrical Engineering.

Electronic Theory
138 pages ▲ Paperback ▲ 6 x 9"
ISBN: 0-7906-1062-0 ▲ Sams: 61062
$16.95 ($22.99 Canada) ▲ March 1995

Industrial Electronics For Technicians
J.A. Sam Wilson & Joseph Risse

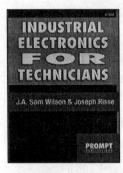

Industrial Electronics for Technicians provides an effective overview of the topics covered in the Industrial Electronics CET test, and is also a valuable reference on industrial electronics in general. This workbench companion book covers the theory and applications of industrial hardware from the technician's perspective, giving students the explanations they need to understand all of the areas required to qualify for CET accreditation. There is simply not a more complete and focused review manual for technicians available on the market today.

J.A. Sam Wilson has written numerous books covering all aspects of the electronics field, and has served as the Director of Technical Publications for NESDA. Joseph Risse develops courses and laboratory experiments in self-study and industrial electronics for International Correspondence Schools and other independent study schools.

Electronic Theory
352 pages ▲ Paperback ▲ 6 x 9"
ISBN: 0-7906-1058-2 ▲ Sams: 61058
$16.95 ($22.99 Canada) ▲ September 1994

Schematic Diagrams
J. Richard Johnson

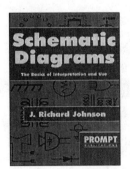

Step by step, *Schematic Diagrams* shows the reader how to recognize schematic symbols and determine their uses and functions in diagrams. Readers will also learn how to design, maintain, and repair electronic equipment as this book takes them logically through the fundamentals of schematic diagrams. Subjects covered include component symbols and diagram formation, functional sequence and block diagrams, power supplies, audio system diagrams, interpreting television receiver diagrams, and computer diagrams. *Schematic Diagrams* is an invaluable instructional tool for students and hobbyists, and an excellent guide for technicians.

J. Richard Johnson has written numerous books covering all aspects of electronics. He worked for many years as an engineer for Bell Laboratories, and was also the managing editor of *Radio Maintenance* magazine. He currently works as a free-lance writer within the electronics field.

Electronic Theory
196 pages ▲ Paperback ▲ 6 x 9"
ISBN: 0-7906-1059-0 ▲ Sams: 61059
$16.95 ($22.99 Canada) ▲ October 1994

CALL 1-800-428-7267 TODAY FOR THE NAME OF YOUR NEAREST PROMPT PUBLICATIONS DISTRIBUTOR

Basic Electricity

Van Valkenburgh, Nooger & Neville, Inc.

Electricity is vital to our everyday lives, yet very few people understand the basic concepts and principles behind it. *Basic Electricity* will provide that understanding to students, laymen, and technicians. From a simplified explanation of the electron to AC/DC machinery, alternators, and other advanced topics, this book is the complete course for mastering the fundamentals of electricity.

The authors have provided a clear understanding of how electricity is produced, measured, controlled, and used. A minimum of mathematics is used for direct explanations of primary cells, magnetism, Ohm's Law, capacitance, transformers, DC generators, and AC motors. Other essential topics covered include conductance, current flow, electromagnetism, and meters. The comprehensive and cumulative index will further help readers in finding subjects of interest quickly, easily, and accurately.

Electronic Theory
736 pages ▲ Paperback ▲ 6 x 9"
ISBN: 0-7906-1041-8 ▲ Sams: 61041
$19.95 ($26.99 Canada) ▲ February 1993

Basic Electricity & DC Circuits

Charles W. Dale, Ed.D

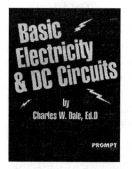

Electricity is constantly at work around your home and community, lighting rooms, running manufacturing facilities, cooling stores and offices, playing radios and stereos, and computing bank accounts. Now you can learn the basic concepts and fundamentals behind electricity, and how it is used and controlled.

Basic Electricity & DC Circuits is designed primarily for the entry-level student. No sophisticated math background or previous knowledge of electricity is assumed, and no matter their background, readers can learn the basic concepts that have enabled mankind to harness and control electricity. The text is arranged to let the readers progress at their own pace, with concepts and terms being introduced as needed for comprehension. Some of the concepts covered include voltage, current, circuits, scientific notation, Ohm's law, DC circuit analysis, capacitors, the RC time constant, and inductors.

Electronic Theory
928 pages ▲ Paperback ▲ 6 x 9"
ISBN: 0-7906-1072-8 ▲ Sams: 61072
$34.95 ($50.95 Canada) ▲ September 1995

Basic Solid-State Electronics

Van Valkenburgh, Nooger & Neville, Inc.

The age of solid-state electronics really began in the late 1950s when solid-state devices started to become available commercially. These solid-state devices were much smaller than vacuum tubes, and used much less power. New circuits were developed to be used in place of existing vacuum-tube circuits, and while the size and weight of electronic equipment decreased rapidly, the reliability increased enormously.

Modern electronics technology manages all aspects of information including generation, transmission, reception, storage, retrieval, manipulation, display, and control. A continuation of the instruction provided in *Basic Electricity*, *Basic Solid-State Electronics* provides the reader with a progressive understanding of the elements that form various electronic systems. Electronic fundamentals covered in the illustrated, easy-to-understand text include semiconductors, power supplies, audio and video amplifiers, transmitters, receivers, and more.

Electronic Theory
944 pages ▲ Paperback ▲ 6 x 9"
ISBN: 0-7906-1042-6 ▲ Sams: 61042
$24.95 ($33.95 Canada) ▲ February 1993

**CALL 1-800-428-7267 TODAY FOR THE NAME OF
YOUR NEAREST PROMPT PUBLICATIONS DISTRIBUTOR**

Semiconductor Essentials
Stephen Kamichik

Readers will gain hands-on knowledge of semiconductor diodes and transistors with help from the information in this book. *Semiconductor Essentials* is a first course in electronics at the technical and engineering levels. Each chapter is a lesson in electronics, with problems included to test understanding of the material presented. This generously illustrated manual is a useful instructional tool for the student and hobbyist, as well as a practical review for professional technicians and engineers. The comprehensive coverage includes semiconductor chemistry, rectifier diodes, zener diodes, transistor biasing, and more.

Author Stephen Kamichik is an electronics consultant who has developed dozens of electronic products and received patents in both the U.S. and Canada. He holds degrees in electrical engineering, and was employed by SPAR, where he worked on the initial prototyping of the Canadarm.

Electronic Theory
112 pages ▲ Paperback ▲ 6 x 9"
ISBN: 0-7906-1071-X ▲ Sams: 61071
$16.95 ($22.99 Canada) ▲ September 1995

Introduction to Microprocessor Theory & Operation
J.A. Sam Wilson & Joseph Risse

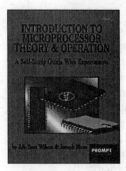

This book takes readers into the heart of computerized equipment and reveals how microprocessors work. By covering digital circuits in addition to microprocessors and providing self-tests and experiments, *Introduction to Microprocessor Theory & Operation* makes it easy to learn microprocessor systems. The text is fully illustrated with circuits, specifications, and pinouts to guide beginners through the ins-and-outs of microprocessors, as well as provide experienced technicians with a valuable reference and refresher tool.

J.A. Sam Wilson has written numerous books covering all aspects of the electronics field, and has served as the Director of Technical Publications for NESDA. Joseph Risse develops courses and laboratory experiments in self-study and industrial electronics for International Correspondence Schools and other independent study schools.

Electronic Theory
211 pages ▲ Paperback ▲ 6 x 9"
ISBN: 0-7906-1064-7 ▲ Sams: 61064
$16.95 ($22.99 Canada) ▲ February 1995

Basic Principles of Semiconductors
Irving M. Gottlieb

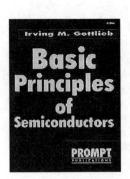

Despite their ever-growing prominence in the electronics industry, semiconductors are still plagued by a stigma which defines them merely as poor conductors. This narrow-sighted view fails to take into account the fact that semiconductors are truly unique alloys whose conductivity is enhanced tenfold by the addition of even the smallest amount of light, voltage, heat, or certain other substances. *Basic Principles of Semiconductors* explores the world of semiconductors, beginning with an introduction to atomic physics before moving onto the structure, theory, applications, and future of these still-evolving alloys. Such a theme makes this book useful to a wide spectrum of practitioners, from the hobbyist and student, right up to the technician and the professional electrician. Irving M. Gottlieb is the author of over ten books in the electrical and electronics fields. *Basic Principles of Semiconductors* is his latest offering, however *Test Procedures for Basic Electronics* is also available from PROMPT® Publications.

Electronic Theory
158 pages ▲ Paperback ▲ 6 x 9"
ISBN: 0-7906-1066-3 ▲ Sams: 61066
$14.95 ($20.95 Canada) ▲ April 1995

CALL 1-800-428-7267 TODAY FOR THE NAME OF YOUR NEAREST PROMPT PUBLICATIONS DISTRIBUTOR

PROMPT®
PUBLICATIONS

Internet Guide to the Electronics Industry

John Adams

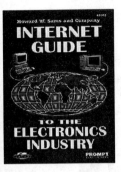

Although the Internet pervades our lives, it would not have been possible without the growth of electronics. It is very fitting then that technical subjects, data sheets, parts houses, and of course manufacturers, are developing new and innovative ways to ride along the Information Superhighway. Whether it's programs that calculate Ohm's Law or a schematic of a satellite system, electronics hobbyists and technicians can find a wealth of knowledge and information on the Internet.

In fact, soon electronics hobbyists and professionals will be able to access on-line catalogs from manufacturers and distributors all over the world, and then order parts, schematics, and other merchandise without leaving home. The *Internet Guide to the Electronics Industry* serves mainly as a directory to the resources available to electronics professionals and hobbyists.

Internet
192 pages ▲ Paperback ▲ 5-1/2 x 8-1/2"
ISBN: 0-7906-1092-2 ▲ Sams: 61092
$16.95 ($22.99 Canada) ▲ December 1996

Real-World Interfacing with Your PC

James "J.J." Barbarello

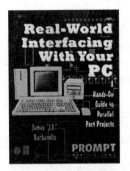

As the computer becomes increasingly prevalent in society, its functions and applications continue to expand. Modern software allows users to do everything from balance a checkbook to create a family tree. Interfacing, however, is truly the wave of the future for those who want to use their computer for things other than manipulating text, data, and graphics.

Real-World Interfacing With Your PC provides all the information necessary to use a PC's parallel port as a gateway to electronic interfacing. In addition to hardware fundamentals, this book provides a basic understanding of how to write software to control hardware.

While the book is geared toward electronics hobbyists, it includes a chapter on project design and construction techniques, a checklist for easy reference, and a recommended inventory of starter electronic parts to which readers at every level can relate.

Computer Technology
119 pages ▲ Paperback ▲ 7-3/8 x 9-1/4"
ISBN: 0-7906-1078-7 ▲ Sams: 61078
$16.95 ($22.99 Canada) ▲ March 1996

ES&T Presents Computer Troubleshooting & Repair

Electronic Servicing & Technology

ES&T is the nation's most popular magazine for professionals who service consumer electronics equipment. PROMPT® Publications, a rising star in the technical publishing business, is combining its publishing expertise with the experience and knowledge of *ES&T's* best writers to produce a new line of troubleshooting and repair books for the electronics market. Compiled from articles and prefaced by the editor in chief, Nils Conrad Persson, these books provide valuable, hands-on information for anyone interested in electronics and product repair.

Computer Troubleshooting & Repair is the second book in the series and features information on repairing Macintosh computers, a CD-ROM primer, and a color monitor. Also included are hard drive troubleshooting and repair tips, computer diagnostic software, networking basics, preventative maintenance for computers, upgrading, and much more.

Computer Technology
288 pages ▲ Paperback ▲ 6 x 9"
ISBN: 0-7906-1087-6 ▲ Sams: 61087
$18.95 ($26.50 Canada) ▲ February 1997

**CALL 1-800-428-7267 TODAY FOR THE NAME OF
YOUR NEAREST PROMPT PUBLICATIONS DISTRIBUTOR**

The Phone Book

Gerald Luecke & James Allen

This book is an installation guide for telephones and telephone accessories. It was written to make it easier for the inexperienced person to install telephones, whether existing ones are being replaced or moved or new ones added, without the hassle and expense of contracting a serviceman. *The Phone Book* begins by explaining the telephone system and its operation, before moving onto clear step-by-step instructions for replacing and adding telephones. With this book, a minimum of tools available around the house, and readily available parts, readers will be able to handle any telephone installation in the home, apartment, or small business.

Gerald Luecke has written articles on integrated circuits and digital technology for numerous trade and professional organizations. James Allen is the President, CEO, and a director of Master Publishing.

Communication
176 pages ▲ Paperback ▲ 7-3/8 x 9-1/4"
ISBN: 0-7906-1028-0 ▲ Sams: 61028
$16.95 ($22.99 Canada) ▲ October 1992

Digital Electronics

Stephen Kamichik

Although the field of digital electronics emerged years ago, there has never been a definitive guide to its theories, principles, and practices — until now. *Digital Electronics* is written as a textbook for a first course in digital electronics, but its applications are varied.

Useful as a guide for independent study, the book also serves as a review for practicing technicians and engineers. And because *Digital Electronics* does not assume prior knowledge of the field, the hobbyist can gain insight about digital electronics.

Some of the topics covered include analog circuits, logic gates, flip-flops, and counters. In addition, a problem set appears at the end of each chapter to test the reader's understanding and comprehension of the materials presented. Detailed instructions are provided so that the readers can build the circuits described in this book to verify their operation.

Electronic Theory
150 pages ▲ Paperback ▲ 7-3/8 x 9-1/4"
ISBN: 0-7906-1075-2 ▲ Sams: 61075
$16.95 ($22.99 Canada) ▲ February 1996

The Right Antenna

Alvis J. Evans

The Right Antenna is intended to provide easy-to-understand information on a wide variety of antennas. It begins by explaining how antennas work and then isolates antennas for TV and FM. A separate chapter is devoted to satellite TV antennas, noise and interference, and antennas used by hams for antenna band operation. The basic concepts of cellular telephone system operation are explained and the most popular antennas are discussed. After studying this book, the reader will be able to select an antenna, place it correctly, and install it properly to obtain maximum performance whether in a strong signal area or in a fringe area.

Alvis Evans is the author of many books on the subject of electricity and electronics for beginning hobbyists and advanced technicians. He teaches seminars and workshops worldwide to members of the trade, as well as being an Associate Professor of Electronics at Tarrant County Junior College.

Communication
112 pages ▲ Paperback ▲ 6 x 9"
ISBN: 0-7906-1022-1 ▲ Sams: 61022
$10.95 ($14.95 Canada) ▲ November 1992

**CALL 1-800-428-7267 TODAY FOR THE NAME OF
YOUR NEAREST PROMPT PUBLICATIONS DISTRIBUTOR**

PROMPT®
PUBLICATIONS

Semiconductor Cross Reference Book
Fourth Edition
Howard W. Sams & Company

This newly revised and updated reference book is the most comprehensive guide to replacement data available for engineers, technicians, and those who work with semiconductors. With more than 490,000 part numbers, type numbers, and other identifying numbers listed, technicians will have no problem locating the replacement or substitution information needed. There is not another book on the market that can rival the breadth and reliability of information available in the fourth edition of the *Semiconductor Cross Reference Book*.

Professional Reference
688 pages ▲ Paperback ▲ 8-1/2 x 11"
ISBN: 0-7906-1080-9 ▲ Sams: 61080
$24.95 ($33.95 Canada) ▲ August 1996

The Component Identifier
and Source Book
Victor Meeldijk

Because interface designs are often reverse engineered using component data or block diagrams that list only part numbers, technicians are often forced to search for replacement parts armed only with manufacturer logos and part numbers.

This source book was written to assist technicians and system designers in identifying components from prefixes and logos, as well as find sources for various types of microcircuits and other components. There is not another book on the market that lists as many manufacturers of such diverse electronic components.

Professional Reference
384 pages ▲ Paperback ▲ 8-1/2 x 11"
ISBN: 0-7906-1088-4 ▲ Sams: 61088
$24.95 ($33.95 Canada) ▲ November 1996

IC Cross Reference Book
Second Edition
Howard W. Sams & Company

The engineering staff of Howard W. Sams & Company assembled the *IC Cross Reference Book* to help readers find replacements or substitutions for more than 35,000 ICs and modules. It is an easy-to-use cross reference guide and includes part numbers for the United States, Europe, and the Far East. This reference book was compiled from manufacturers' data and from the analysis of consumer electronics devices for PHOTOFACT® service data, which has been relied upon since 1946 by service technicians worldwide.

Professional Reference
192 pages ▲ Paperback ▲ 8-1/2 x 11"
ISBN: 0-7906-1096-5 ▲ Sams: 61096
$19.95 ($26.99 Canada) ▲ November 1996

Tube Substitution
Handbook
William Smith & Barry Buchanan

The most accurate, up-to-date guide available, the *Tube Substitution Handbook* is useful to antique radio buffs, old car enthusiasts, and collectors of vintage ham radio equipment. In addition, marine operators, microwave repair technicians, and TV and radio technicians will find the *Handbook* to be an invaluable reference tool.

The *Tube Substitution Handbook* is divided into three sections, each preceded by specific instructions. These sections are vacuum tubes, picture tubes, and tube basing diagrams.

Professional Reference
149 pages ▲ Paperback ▲ 6 x 9"
ISBN: 0-7906-1036-1 ▲ Sams: 61036
$16.95 ($22.99 Canada) ▲ March 1995

**CALL 1-800-428-7267 TODAY FOR THE NAME OF
YOUR NEAREST PROMPT PUBLICATIONS DISTRIBUTOR**

Alternative Energy
Mark E. Hazen

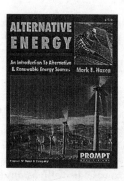

This book is designed to introduce readers to the many different forms of energy mankind has learned to put to use. Generally, energy sources are harnessed for the purpose of producing electricity. This process relies on transducers to transform energy from one form into another. *Alternative Energy* will not only address transducers and the five most common sources of energy that can be converted to electricity, it will also explore solar energy, the harnessing of the wind for energy, geothermal energy, and nuclear energy.

This book is designed to be an introduction to energy and alternate sources of electricity. Each of the nine chapters are followed by questions to test comprehension, making it ideal for students and teachers alike. In addition, listings of World Wide Web sites are included so that readers can learn more about alternative energy and the organizations devoted to it.

Professional Reference
320 pages ▲ Paperback ▲ 7-3/8 x 9-1/4”
ISBN: 0-7906-1079-5 ▲ Sams: 61079
$18.95 ($25.95 Canada) ▲ October 1996

The Complete RF Technician's Handbook
Cotter W. Sayre

The *Complete RF Technician's Handbook* will furnish the working technician or student with a solid grounding in the latest methods and circuits employed in today's RF communications gear. It will also give readers the ability to test and troubleshoot transmitters, transceivers, and receivers with absolute confidence. Some of the topics covered include reactance, phase angle, logarithms, diodes, passive filters, amplifiers, and distortion. Various multiplexing methods and data, satellite, spread spectrum, cellular, and microwave communication technologies are discussed.

Cotter W. Sayre is an electronics design engineer with Goldstar Development, Inc., in Lake Elsinore, California. He is a graduate of Los Angeles Pierce College and is certified by the National Association of Radio and Telecommunications Engineers, as well as the International Society of Electronics Technicians.

Professional Reference
281 pages ▲ Paperback ▲ 8-1/2 x 11”
ISBN: 0-7906-1085-X ▲ Sams: 61085
$24.95 ($33.95 Canada) ▲ July 1996

Surface-Mount Technology for PC Boards
James K. Hollomon, Jr.

The race to adopt surface-mount technology, or SMT as it is known, has been described as the latest revolution in electronics. This book is intended for the working engineer or manager, the student or the interested layman, who would like to learn to deal effectively with the many trade-offs required to produce high manufacturing yields, low test costs, and manufacturable designs using SMT. The valuable information presented in *Surface-Mount Technology for PC Boards* includes the benefits and limitations of SMT, SMT and FPT components, manufacturing methods, reliability and quality assurance, and practical applications.

James K. Hollomon, Jr. is the founder and president of AMTI, an R&D and prototyping service concentrating on miniaturization and low-noise, high-speed applications. He has nearly 20 years experience in engineering, marketing, and managing firms dealing with leadless components. His previous appointments include national president of the Surface-Mount Technology Association.

Professional Reference
510 pages ▲ Paperback ▲ 7 x 10”
ISBN: 0-7906-1060-4 ▲ Sams: 61060
$26.95 ($36.95 Canada) ▲ July 1995

CALL 1-800-428-7267 TODAY FOR THE NAME OF
YOUR NEAREST PROMPT PUBLICATIONS DISTRIBUTOR

Is This Thing On?

Gordon McComb

Is This Thing On? takes readers through each step of selecting components, installing, adjusting, and maintaining a sound system for small meeting rooms, churches, lecture halls, public-address systems for schools or offices, or any other large room.

In easy-to-understand terms, drawings and illustrations, *Is This Thing On?* explains the exact procedures behind connections and troubleshooting diagnostics. With the help of this book, hobbyists and technicians can avoid problems that often occur while setting up sound systems for events and lectures.

Is This Thing On? covers basic components of sound systems, the science of acoustics, enclosed room, sound system specifications, wiring sound systems, and how to install wireless microphones, CD players, portable public-address systems, and more.

Audio Technology
136 pages ▲ Paperback ▲ 6 x 9"
ISBN: 0-7906-1081-7 ▲ Sams: 61081
$14.95 ($20.95 Canada) ▲ April 1996

Advanced Speaker Designs

Ray Alden

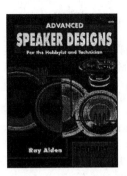

Advanced Speaker Designs shows the hobbyist and the experienced technician how to create high-quality speaker systems for the home, office, or auditorium. Every part of the system is covered in detail, from the driver and crossover network to the enclosure itself. Readers can build speaker systems from the parts lists and instructions provided, or they can actually learn to calculate design parameters, system responses, and component values with scientific calculators or PC software.

This book includes construction plans for seven complete systems, easy-to-understand instructions and illustrations, and chapters on sealed and vented enclosures. There is also emphasis placed on enhanced bass response, computer-aided speaker design, and driver parameters. *Advanced Speaker Designs* is a companion book to *Speakers for Your Home and Automobile*, also available from Prompt® Publications.

Audio Technology
136 pages ▲ Paperback ▲ 6 x 9"
ISBN: 0-7906-1070-1 ▲ Sams: 61070
$16.95 ($22.99 Canada) ▲ July 1995

Making Sense of Sound

Alvis J. Evans

This book deals with the subject of sound — how it is detected and processed using electronics in equipment that spans the full spectrum of consumer electronics. It concentrates on explaining basic concepts and fundamentals to provide easy-to-understand information, yet it contains enough detail to be of high interest to the serious practitioner. Discussion begins with how sound propagates and common sound characteristics, before moving on to the more advanced concepts of amplification and distortion. *Making Sense of Sound* was designed to cover a broad scope, yet in enough detail to be a useful reference for readers at every level.

Alvis Evans is the author of many books on the subject of electricity and electronics for beginning hobbyists and advanced technicians. He teaches seminars and workshops worldwide to members of the trade, as well as being an Associate Professor of Electronics at Tarrant County Junior College.

Audio Technology
112 pages ▲ Paperback ▲ 6 x 9"
ISBN: 0-7906-1026-4 ▲ Sams: 61026
$10.95 ($14.95 Canada) ▲ November 1992

CALL 1-800-428-7267 TODAY FOR THE NAME OF YOUR NEAREST PROMPT PUBLICATIONS DISTRIBUTOR

Speakers for Your Home & Automobile

Gordon McComb, Alvis J. Evans, & Eric J. Evans

The cleanest CD sound, the quietest turntable, or the clearest FM signal are useless without a fine speaker system. This book not only tells readers how to build quality speaker systems, it also shows them what components to choose and why. The comprehensive coverage includes speakers, finishing touches, construction techniques, wiring speakers, and automotive sound systems.

Gordon McComb has written over 35 books and 1,000 magazine articles which have appeared in such publications as *Popular Science, Video, Omni, Popular Electronics,* and *PC World*. His writings has spanned a wide range of subjects, from computers, to video, to robots. Alvis and Eric Evans are the co-authors of many books and articles on the subject of electricity and electronics. Alvis is also an Associate Professor of Electronics at Tarrant County Junior College in Ft. Worth, Texas.

Audio Technology
164 pages ▲ Paperback ▲ 6 x 9"
ISBN: 0-7906-1025-6 ▲ Sams: 61025
$14.95 ($20.95 Canada) ▲ November 1992

Sound Systems for Your Automobile

Alvis J. Evans & Eric J. Evans

This book provides the average vehicle owner with the information and skills needed to install, upgrade, and design automotive sound systems. From terms and definitions straight up to performance objectives and cutting layouts, *Sound Systems* will show the reader how to build automotive sound systems that provide occupants with live performance reproductions that rival home audio systems.

Whether starting from scratch or upgrading, this book uses easy-to-follow steps to help readers plan their system, choose components and speakers, and install and interconnect them to achieve the best sound quality possible. Installations on specific types of vehicles are discussed, including separate chapters on coupes and sedans, hatchbacks, pick-up trucks, sport utility vehicles, and vans. Alvis J. Evans is the author of many books on the subjects of electricity and electronics for both beginning hobbyists and advanced technicians.

Audio Technology
124 pages ▲ Paperback ▲ 6 x 9"
ISBN: 0-7906-1046-9 ▲ Sams: 61046
$16.95 ($22.99 Canada) ▲ January 1994

Radios of the Baby Boom Era

Howard W. Sams & Company

This is the first comprehensive picture reference to the classic radios of the post-World War II years, when radio blossomed into the nation's personal entertainment accessory. From Howard Sams' PHOTOFACT® service data library come these classic photos to help readers identify and fix their basement treasures or get started on the rewarding hobby of radio collecting. Each volume in the series is arranged by brand name and contains more than 675 photographs of popular radios, including radio/phonograph combinations. Brand names covered include Admiral, Geloso, Motorola, RCA Victor, and Zenith among others. Individual volumes are available for $16.95 each, and the entire set sells for just $96.50.

This book is written by the editors of Howard W. Sams & Company. These same people created the PHOTOFACT® series of repair data, which has grown to become the nation's leading source of after-market service information. The pictures and data in this series are taken from the early sets.

Audio Technology
Six-Volume Set ▲ Paperback ▲ 8-1/2 x 11"
ISBN: 0-7906-1009-4 ▲ Sams: 61009
$96.50 ($133.95 Canada) ▲ May 1991

CALL 1-800-428-7267 TODAY FOR THE NAME OF YOUR NEAREST PROMPT PUBLICATIONS DISTRIBUTOR

▼   ▼   ▼   ▼   ▼   ▼   ▼   ▼   ▼   ▼   ▼   ▼

The Howard W. Sams
Troubleshooting & Repair Guide to TV
Howard W. Sams & Company

The Howard W. Sams Trouble-shooting & Repair Guide to TV is the most complete and up-to-date television repair book available. Included in its more than 300 pages is complete repair information for all makes of TVs, timesaving features that even the pros don't know, comprehensive basic electronics information, and extensive coverage of common TV symptoms.

This repair guide is completely illustrated with useful photos, schematics, graphs, and flowcharts. It covers audio, video, technician safety, test equipment, power supplies, picture-in-picture, and much more. *The Howard W. Sams Troubleshooting & Repair Guide to TV* was written, illustrated, and assembled by the engineers and technicians of Howard W. Sams & Company. This book is the first truly comprehensive television repair guide published in the 90s, and it contains vast amounts of information never printed in book form before.

Video Technology
384 pages ▲ Paperback ▲ 8-1/2 x 11"
ISBN: 0-7906-1077-9 ▲ Sams: 61077
$29.95 ($39.95 Canada) ▲ June 1996

The In-Home VCR Mechanical
Repair & Cleaning Guide
Curt Reeder

Like any machine that is used in the home or office, a VCR requires minimal service to keep it functioning well and for a long time. However, a technical or electrical engineering degree is not required to begin regular maintenance on a VCR. *The In-Home VCR Mechanical Repair & Cleaning Guide* shows readers the tricks and secrets of VCR maintenance using just a few small hand tools, such as tweezers and a power screwdriver.

This book is also geared toward entrepreneurs who may consider starting a new VCR service business of their own. The vast information contained in this guide gives a firm foundation on which to create a personal niche in this unique service business. This book is compiled from the most frequent VCR malfunctions Curt Reeder has encountered in the six years he has operated his in-home VCR repair and cleaning service.

Video Technology
222 pages ▲ Paperback ▲ 8-3/8 x 10-7/8"
ISBN: 0-7906-1076-0 ▲ Sams: 61076
$19.95 ($26.99 Canada) ▲ April 1996

ES&T Presents TV Troubleshooting & Repair
Electronic Servicing & Technology Magazine

TV set servicing has never been easy. The service manager, service technician, and electronics hobbyist need timely, insightful information in order to locate the correct service literature, make a quick diagnosis, obtain the correct replacement components, complete the repair, and get the TV back to the owner.

ES&T Presents TV Troubleshooting & Repair presents information that will make it possible for technicians and electronics hobbyists to service TVs faster, more efficiently, and more economically, thus making it more likely that customers will choose not to discard their faulty products, but to have them restored to service by a trained, competent professional.

Originally published in *Electronic Servicing & Technology*, the chapters in this book are articles written by professional technicians, most of whom service TV sets every day. These chapters provide general descriptions of television circuit operation, detailed service procedures, and diagnostic hints.

Video Technology
226 pages ▲ Paperback ▲ 6 x 9"
ISBN: 0-7906-1086-8 ▲ Sams: 61086
$18.95 ($25.95 Canada) ▲ August 1996

CALL 1-800-428-7267 TODAY FOR THE NAME OF
YOUR NEAREST PROMPT PUBLICATIONS DISTRIBUTOR

Theory & Design of Loudspeaker Enclosures

Dr. J. Ernest Benson

The design of loudspeaker enclosures, particularly vented enclosures, has been a subject of continuing interest since 1930. Since that time, a wide range of interests surrounding loudspeaker enclosures have sprung up that grapple with the various aspects of the subject, especially design. *Theory & Design of Loudspeaker Enclosures* lays the groundwork for readers who want to understand the general functions of loudspeaker enclosure systems and eventually experiment with their own design.

Written for design engineers and technicians, students and intermediate-to-advanced level acoustics enthusiasts, this book presents a general theory of loudspeaker enclosure systems. Full of illustrated and numerical examples, this book examines diverse developments in enclosure design, and studies the various types of enclosures as well as varying parameter values and performance optimization.

Audio Technology
244 pages ▲ Paperback ▲ 6 x 9"
ISBN: 0-7906-1093-0 ▲ Sams: 61093
$19.95 ($26.99 Canada) ▲ August 1996

TV Video Systems

L.W. Pena & Brent A. Pena

Knowing which video programming source to choose, and knowing what to do with it once you have it, can seem overwhelming. Covering standard hard-wired cable, large-dish satellite systems, and DSS, *TV Video Systems* explains the different systems, how they are installed, their advantages and disadvantages, and how to troubleshoot problems. This book presents easy-to-understand information and illustrations covering installation instructions, home options, apartment options, detecting and repairing problems, and more. The in-depth chapters guide you through your TV video project to a successful conclusion.

L.W. Pena is an independent certified cable TV technician with 14 years of experience who has installed thousands of TV video systems in homes and businesses. Brent Pena has eight years of experience in computer science and telecommunications, with additional experience as a cable installer.

Video Technology
124 pages ▲ Paperback ▲ 6 x 9"
ISBN: 0-7906-1082-5 ▲ Sams: 61082
$14.95 ($20.95 Canada) ▲ June 1996

The Video Book

Gordon McComb

Televisions and video cassette recorders have become part of everyday life, but few people know how to get the most out of these home entertainment devices. *The Video Book* offers easy-to-read text and clearly illustrated examples to guide readers through the use, installation, connection, and care of video system components. Simple enough for the new buyer, yet detailed enough to assure proper connection of the units after purchase, this book is a necessary addition to the library of every modern video consumer. Topics included in the coverage are the operating basics of TVs, VCRs, satellite systems, and video cameras; maintenance and troubleshooting; and connectors, cables, and system interconnections.

Gordon McComb has written over 35 books and 1,000 magazine articles, which have appeared in such publications as *Popular Science*, *Video*, *PC World*, and *Omni*, as well as many other top consumer and trade publications. His writing has spanned a wide range of subjects, from computers to video to robots.

Video Technology
192 pages ▲ Paperback ▲ 6 x 9"
ISBN: 0-7906-1030-2 ▲ Sams: 61030
$16.95 ($22.99 Canada) ▲ October 1992

CALL 1-800-428-7267 TODAY FOR THE NAME OF YOUR NEAREST PROMPT PUBLICATIONS DISTRIBUTOR

▼ ▼ ▼ ▼ ▼ ▼ ▼ ▼ ▼ ▼ ▼ ▼

Advanced Electronic Projects
Stephen Kamichik

The projects in this book were designed to yield the ultimate in performance and features. They can be built by the experienced electronic hobbyist as well as by electronic technicians and engineers. An explanation of each circuit is given to enable readers to troubleshoot the project should it not work. *Advanced Electronic Projects* is divided into four parts: active filters, residential surround sound systems, advanced projects for the home, and projects for the car. Readers will gain valuable experience in the field of advanced electronics by learning how to build the interesting and useful projects in this book.

Stephen Kamichik is an electronics consultant who has developed dozens of electronic products and received patents in both the U.S. and Canada. He holds degrees in electrical engineering, and was employed for a number of years at SPAR, where he worked on the prototyping of the Canadarm.

Electronic Project

123 pages ▲ Paperback ▲ 6 x 9"
ISBN: 0-7906-1065-5 ▲ Sams: 61065
$18.95 ($25.95 Canada) ▲ April 1995

ES&T Presents Test Equipment Troubleshooting & Repair
Electronic Servicing & Technology Magazine

ES&T, the nation's most popular magazine for professionals who service consumer electronics equipment, has combined its experience and knowledge with the publishing expertise of PROMPT® to produce this new book. Compiled from articles and prefaced by *ES&T's* editor in chief, Nils Conrad Persson, this book will provide valuable hands-on information for anyone interested in electronics and product repair. *Test Equipment Troubleshooting & Repair* includes the latest information on how to choose the best equipment, how to build test equipment and accessories, how to set up the ideal service bench, and how to put together a practical technician's tool kit. A special chapter also discusses surface mount technology and how to test its components. The list of featured writers includes some of *ES&T's* brightest starts like Homer Davidson, Conrad Persson, Vaughn D. Martin, Brian Phelps, and others.

Test and Measurement

288 pages ▲ Paperback ▲ 6 x 9"
ISBN: 0-7906-1089-2 ▲ Sams: 61089
$18.95 ($25.95 Canada) ▲ March 1997

Electronic Control Projects
Henry C. Smith & Craig B. Foster

The most enjoyable way to learn about electronics and to gain practical experience is by building projects. This book was designed both to demonstrate how and why an electronic circuit works and to apply that knowledge to real-world problems. Written to be as educational as possible, the projects involve the reader in the synthesis of the circuit. Some of the projects featured include an electronic thermostat, a touch-operated switch, and a telephone remote control and call screening device. Each of the projects were built and tested by the authors to ensure that they are functional and dependable.

Henry Smith attended American University where he majored in physics and has a degree in electronics engineering technology from CIE. He is currently a field engineer for the Eastman Kodak Company. Craig Foster is also a field engineer for Kodak, and including his experience in the U.S. Navy, has worked in the electronics field for over 20 years.

Electronic Project

158 pages ▲ Paperback ▲ 6 x 9"
ISBN: 0-7906-1044-2 ▲ Sams: 61044
$16.95 ($22.99 Canada) ▲ November 1993

CALL 1-800-428-7267 TODAY FOR THE NAME OF YOUR NEAREST PROMPT PUBLICATIONS DISTRIBUTOR

PROMPT®
PUBLICATIONS

Test Procedures for Basic Electronics

Irving M. Gottlieb

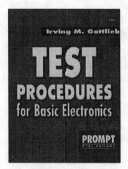

Many useful tests and measurements are covered in this book, and the emphasis is always on the deployment of commonly available instruments, rather than laboratory types. Test procedures and measurements are reinforced by the appropriate basic principles, and examples of test and measurement setups are given to make concepts more practical. Students, hobbyists, and professionals will find the whats and whys of obtaining useful results, whether they are repairing a modern CD player or restoring an antique radio. Other topics covered include DC resistance, voltage and current, and semiconductor devices.

Irving Gottlieb is the author of over ten books on the electrical and electronics fields. *Test Procedures* is the first book he published with PROMPT®, but he has since added *Basic Principles of Semiconductors* to his list of credits. Gottlieb is now traveling the country giving lectures.

Test and Measurement
356 pages ▲ Paperback ▲ 7-3/8 x 9-1/4"
ISBN: 0-7906-1063-9 ▲ Sams: 61063
$16.95 ($22.99 Canada) ▲ December 1994

The Multitester Guide

Alvis J. Evans

Making basic electrical circuit measurements of voltage, current, or resistance with a meter helps significantly in understanding an electrical circuit to tell if it is operating properly of if repair or maintenance is necessary to correct its operation. *The Multitester Guide* provides clearly illustrated, step-by-step instructions so that readers will see the versatility and practicality of the multitester. In addition to the basic functions and uses of multitesters, the easy-to-understand text covers such topics as the measurement of basic electrical components, automotive circuit measurements, and the operation and measurements on typical shop tools.

Alvis J. Evans is an associate professor of electronics at Tarrant County Junior College. As the author of many books on the subject of electricity and electronics for beginning hobbyists as well as technicians, he is in demand to teach seminars and workshops nationwide to members of the trade.

Test and Measurement
156 pages ▲ Paperback ▲ 6 x 9"
ISBN: 0-7906-1027-2 ▲ Sams: 61027
$14.95 ($20.95 Canada) ▲ December 1992

VOM and DVM Multitesters

Alvis J. Evans

This book was written to help readers understand how meters work and how to use them to make basic electrical measurements. It emphasizes "how to do" the various measurements, in addition to providing a general understanding of basic concepts and fundamentals. The book begins by explaining the basic concepts of VOM and DVM, relating how the analog and digital meter differ, and the advantages and disadvantages of each. It then moves into a discussion on multitester measurements, basic DC and AC measurements, voltage, meter loading, reference points, inductors, resistors, current, and resistance measurements.

Alvis J. Evans is an associate professor of electronics at Tarrant County Junior College. As the author of many books on the subject of electricity and electronics for beginning hobbyists as well as technicians, he is in demand to teach seminars and workshops nationwide to members of the trade.

Test and Measurement
144 pages ▲ Paperback ▲ 6 x 9"
ISBN: 0-7906-1031-0 ▲ Sams: 61031
$14.95 ($20.95 Canada) ▲ October 1992

CALL 1-800-428-7267 TODAY FOR THE NAME OF YOUR NEAREST PROMPT PUBLICATIONS DISTRIBUTOR

Managing the Computer Power Environment
Mark Waller

Clean power is what every computer system needs to operate without error. But electricity's voyage from utility company to home or office introduces noise, surges, static, and a host of gremlins that can seriously affect computer performance and data security. Written for data processing specialists, field engineers, technicians, and computer network professionals, *Managing the Computer Power Environment* provides the background in electrical technology that will help readers understand and control the quality of the power that drives their computer system. Covering utility power, grounding, power distribution units, and back-up power systems and conditioners, Mark Waller prepares the reader to manage the demons of electrical destruction through ensuring clean power for your electronic system. Mark Waller is president of the Waller Group, Inc., a company specializing in solving electrical power and grounding problems.

Electrical Technology
174 pages ▲ Paperback ▲ 7-3/8 x 9-1/4"
ISBN: 0-7906-1020-5 ▲ Sams: 61020
$19.95 ($26.99 Canada) ▲ April 1992

Surges, Sags and Spikes
Mark Waller

Surges, sags, spikes, brownouts, blackouts, lightning and other damaging electrical power disturbances can render a personal computer system and its data useless in a few milliseconds — unless you're prepared. Mark Waller's *Surges, Sags and Spikes* is written for all personal computer users concerned with protecting their computer systems against a hostile electrical environment. In easy-to-understand, nontechnical language, the author takes a comprehensive look at approaches to solving computer power problems. Helpful diagrams and photographs are included to document computer power needs and solutions.

Mark Waller is an award-winning author whose numerous articles have appeared in such magazines as *Byte, Datamation,* and *Network World.* He is the author of another book dealing with power entitled *Managing the Computer Power Environment,* also available from PROMPT® Publications.

Electrical Technology
220 pages ▲ Paperback ▲ 7-3/8 x 9-1/4"
ISBN: 0-7906-1019-1 ▲ Sams: 61019
$19.95 ($26.99 Canada) ▲ April 1992

Harmonics
Mark Waller

Harmonics is the essential guide to understanding all of the issues and areas of concern surrounding harmonics and the recognized methods for dealing with them. Covering nonlinear loads, multiple PCs, K-factor transformers, and more, Mark Waller prepares the reader to manage problems often encountered in electrical distribution systems that can be solved easily through an understanding of harmonics, current, and voltage. This book is a useful tool for system and building engineers, electricians, maintenance personnel, and all others concerned about protecting and maintaining the quality of electrical power systems.

Mark Waller is president of the Waller Group, Inc., and specializes in harmonic analysis and in solving electrical power and grounding problems for facilities. He has been actively involved in the field of electrical power quality for many years. Waller has a broad background in all aspects of power quality, power protection, and system integrity.

Electrical Technology
132 pages ▲ Paperback ▲ 7-3/8 x 9-1/4"
ISBN: 0-7906-1048-5 ▲ Sams: 61048
$24.95 ($33.95 Canada) ▲ May 1994

CALL 1-800-428-7267 TODAY FOR THE NAME OF YOUR NEAREST PROMPT PUBLICATIONS DISTRIBUTOR

Electrical Wiring
Arthur C. Seale, Jr.

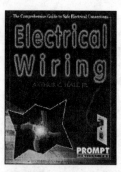

An excellent training guide for electrical trainees, and a valuable reference guide for technicians and hobbyists, *Electrical Wiring* provides the reader with expert advice on how to safely complete electrical wiring connections. Complete with step-by-step instructions and clearly illustrated examples, this easy-to-understand introductory guide is designed to be the perfect instructional tool for anyone interested in learning about electrical wiring and connections. *Electrical Wiring* covers low-voltage wiring applications, wiring for residential constructions, protection methods for short circuits and overloads, and branch circuits inside the home.

Arthur C. Seale, Jr. was an electrical engineer for Long Island Lighting Company until his retirement. He directed installation and maintenance operations of several electrical divisions during his 40-year tenure. This book represents Mr. Seale's professional and personal devotion to the field.

Electrical Technology
274 pages ▲ Paperback ▲ 7-3/8 x 9-1/4"
ISBN: 0-7906-1073-6 ▲ Sams: 61073
$19.95 ($26.99 Canada) ▲ January 1996

Security Systems
Gordon McComb

Security Systems is about making homes safer places to live and protecting cars from vandals and thieves. It is not only a buyer's guide to help readers select the right kind of alarm system for their home and auto, it also shows them how to install the various components. Learning to design, install, and use alarm systems saves a great deal of money, but it also allows people to learn the ins and outs of the system so that it can be used more effectively. This book is divided into eight chapters, including home security basics, warning devices, sensors, control units, remote paging automotive systems, and case histories.

Gordon McComb has written over 35 books and 1,000 magazine articles which have appeared in such publications as *Omni* and *PC World*. In addition, he is the coauthor of PROMPT® Publication's *Speakers for Your Home and Auto*. His writing has spanned a wide range of subjects.

Electronic Project
130 pages ▲ Paperback ▲ 6 x 9"
ISBN: 0-7906-1054-X ▲ Sams: 61054
$16.95 ($22.99 Canada) ▲ July 1994

The Electricity Book
Gene McWhorter

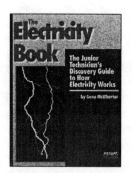

The Electricity Book is the junior technician's discovery guide to how electricity works. It is primarily designed to expand the knowledge of science fair project participants, as it will help them to understand the underlying principles of electricity when they are used in a project. This book will also be of use to instructors and students in eighth grade and beginning high school science classes, and technology classes, hobbyists, and beginning electricians. Some of the topics covered include direct-current circuits, electric charge, electrons, atoms, and resistance. Multiple choice quizzes follow each chapter to help reinforce the learning.

Gene McWhorter is a free-lance writer in the fields of industrial processing, electronics, and chemical processing. His work has been published in *Scientific American* magazine among others, and his other books include *Understanding Digital Electronics* and *Understanding Solid-State Electronics*.

Electrical Technology
128 pages ▲ Paperback ▲ 8-1/2 x 11"
ISBN: 0-7906-1023-X ▲ Sams: 61023
$14.95 ($20.95 Canada) ▲ December 1992

CALL 1-800-428-7267 TODAY FOR THE NAME OF YOUR NEAREST PROMPT PUBLICATIONS DISTRIBUTOR

About The Author

Vaughn D. Martin is a senior electrical engineer with the Department of the Air Force. Previously he worked at Magnavox and ITT Aerospace/Optics, where he acquired his fascination with optoelectronics. He has published numerous articles in trade, amateur radio, electronic hobbyist, troubleshooting and repair, and optoelectronics magazines. He has also written several books covering a wide range of topics in the field of electronics.

COMING SUMMER 1997!
Optoelectronics
Volume 3
Vaughn D. Martin

Optoelectronics, Volume 3 is an Advanced self-teaching text that contains:

- Information on Fiber Optics!

- Lab experiments for your own home lab, professional lab, or classroom!

- AND MUCH MORE!

Optoelectronics, Volume 3
400 pages ♦ Paperback ♦ 8-1/2 x 11"
ISBN: PENDING ♦ Sams: PENDING
$29.95 ♦ August 1997

PROMPT® Publications was created in 1991, and has grown to become one of the top technical imprints in the nation. Concentrating its efforts on technical books designed both for the novice and the experienced electronics technician, PROMPT® published more than 70 books in its first six years, with another 30 scheduled to go to press this year. Each and every PROMPT® book provides a clear understanding of the principles involved in the installation, maintenance, and performance of the electronic devices that have become such a large part of our everyday lives.

Be sure to visit Sams online at **http://www.hwsams.com**. This fully interactive site provides guests with complete book summaries, new release dates, product updates, and order information. Stay up-to-date on the newest and the best from Howard Sams, your technology connection to the future.

A Division of Howard W. Sams & Company\
A Bell ATlantic Company
Indianapo,is,

**CALL 1-800-428-7267 TODAY FOR THE NAME OF
YOUR NEAREST PROMPT PUBLICATIONS DISTRIBUTOR**

COMING IN 1997
FROM PROMPT PUBLICATIONS

The Professional Macintosh Repair Guide
By David Presnell

COMING SPRING 1997

A step-by-step guide to professional Macintosh computer repair. Covers virtually all of the hardware and software problems you are likely to encounter with your Macintosh!

The Professional Macintosh Repair Guide
400 pages ◆ Paperback ◆ 8-1/2 x 11"
ISBN: 0-7906-1090-6 ◆ Sams: 61090
MAY 1997
$29.95

Build Your Own Home Lab
By Clement Pepper

COMING SUMMER 1997

Learn about everything you need to construct an electronics lab in your own home! Contains tips on equipment, work areas, furniture, electronics needs, and much, much more.

Build Your Own Home Lab
224 pages ◆ Paperback ◆ 7-3/8 x 9-1/4"
ISBN: 0-7906-1108-2 ◆ Sams: 61108
JULY 1997
$19.95

Electronic Servicing Techniques
by Sam Wilson & Joe Risse

COMING FALL 1997

A general reference guide that covers basic and advanced techniques for troubleshooting various electronic devices and components, this book places special emphasis on the growing field of digital electronics!

Electronic Servicing Techniques
352 pages ◆ Paperback ◆ 7-3/8 x 9-1/4"
ISBN: 0-7906-1107-4 ◆ Sams: 61107
OCTOBER 1997
$24.95

CALL 1-800-428-7267 TODAY FOR THE NAME OF YOUR NEAREST PROMPT PUBLICATIONS DISTRIBUTOR

COMING IN 1997
FROM PROMPT PUBLICATIONS

The Howard W. Sams Computer Monitor Troubleshooting & Repair Guide
by Joe Desposito

COMING SUMMER 1997

Learn how to troubleshoot computer monitor problems with the information contained in this book!

The Howard W. Sams' Computer Monitor Troubleshooting & Repair Guide
336 pages ✦ Paperback ✦ 8-1/2 x 11"
ISBN: 0-7906-1100-7 ✦ Sams: 61100
JUNE 1997
$29.95

The Howard W. Sams and Company Road Map to Success

COMING SUMMER 1997

Want to start your own electronics business? This book will show you how already-established businesses and professionals got started, and how to use their experiences to help get your own business going!

The Howard W. Sams and Company Road Map to Success
352 pages ✦ Paperback ✦ 8-1/2 x 11"
ISBN: 0-7906-1101-5 ✦ Sams: 61101
JULY 1997
$24.95

The Howard W. Sams Camcorder Troubleshooting & Repair Guide
by Joe Desposito

COMING FALL 1997

Learn how to prevent camcorder problems, and how to troubleshoot & repair problems when they arise! Covers models up to the most recent.

The Howard W. Sams Camcorder Troubleshooting & Repair Guide
336 pages ✦ Paperback ✦ 8-1/2 x 11"
ISBN: 0-7906-1105-8 ✦ Sams: 61105
NOVEMBER 1997
$29.95

**CALL 1-800-428-7267 TODAY FOR THE NAME OF
YOUR NEAREST PROMPT PUBLICATIONS DISTRIBUTOR**

COMING IN 1997
FROM PROMPT PUBLICATIONS

ES&T Presents
Computer Troubleshooting & Repair

COMING WINTER 1997

Continuing the series introduced with *ES&T Presents TV Troubleshooting & Repair*, this volume contains useful information written by professional technicians for *Electronic Servicing &Technology Magazine*!

ES&T Presents Computer Troubleshooting & Repair
288 pages ✦ Paperback ✦ 6 x 9"
ISBN: 0-7906-1087-6 ✦ Sams: 61087
FEBRUARY 1997
$18.95

ES&T Presents
The Test Equipment Guide

COMING SPRING 1997

ES&T Presents The Test Equipment Guide presents you with useful and informative tips about test equipment and how to get the best possible test results. Contains the best articles on the subject originally printed in *ES&T Magazine*!

ES&T Presents The Test Equipment Guide
288 pages ✦ Paperback ✦ 6 x 9"
ISBN: 0-7906-1089-2 ✦ Sams: 61089
MARCH 1997
$18.95

Electronic Projects for the 21st Century
by John Iovine

COMING SUMMER 1997

Get a jump on 21st Century technology with the projects presented in this book! Presents interesting and challenging electronic projects designed in anticipation of the electronics devices and technologies newly emerging, in time for the beginning of the next millennia!

Electronic Projects for the 21st Century
256 pages ✦ Paperback ✦ 7-3/8 x 9-1/4"
ISBN: 0-7906-1103-1 ✦ Sams: 61103
JULY 1997
$19.95

CALL 1-800-428-7267 TODAY FOR THE NAME OF
YOUR NEAREST PROMPT PUBLICATIONS DISTRIBUTOR

COMING IN 1997
FROM PROMPT PUBLICATIONS

PC Hardware Projects, Volume 1
by James "J.J." Barbarello

COMING WINTER 1997

From the author of *Real-World Interfacing With Your PC* comes a series of books dealing with more hands-on hardware projects and applications for your personal computer!

PC Hardware Projects, Volume 1
256 pages ♦ Paperback ♦ 7-3/8 x 9-1/4"
ISBN: 0-7906-1104-X ♦ Sams: 61104
FEBRUARY 1997
$24.95

PC Hardware Projects, Volume 2
by James "J.J." Barbarello

COMING SUMMER 1997

Continuing the projects introduced in Volume 1, *PC Hardware Projects, Volume 2* presents more advanced PC projects that you can implement in your own home!

PC Hardware Projects, Volume 1
256 pages ♦ Paperback ♦ 7-3/8 x 9-1/4"
ISBN: 0-7906-1109-0 ♦ Sams: 61109
JULY 1997
$24.95

The Howard W. Sams Complete Guide to VCR Troubleshooting & Repair
by Joe Desposito

COMING SPRING 1997

Using documentation and schematics assembled by Howard W. Sams' own technicians and engineers, this book tells you what you need to know to maintain and repair your VCR in a comprehensive and easy-to-follow manner. Save money on repairs with the information presented in this book!

The Howard W. Sams Complete Guide to VCR Troubleshooting & Repair
336 pages ♦ Paperback ♦ 8-1/2 x 11"
ISBN: 0-7906-1102-3 ♦ Sams: 61102
MARCH 1997
$29.95

**CALL 1-800-428-7267 TODAY FOR THE NAME OF
YOUR NEAREST PROMPT PUBLICATIONS DISTRIBUTOR**

PROMPT
PUBLICATIONS